Comprehensive™

BIOTECHNOLOGY–I

Comprehensive™

BIOTECHNOLOGY–I

Cell Biology and Genetics

Dr. V. Sreekrishna

Prof. Post Graduate

Deptt. of Biotechnology

Dayanand Sagar Educational Institute Bangalore

NEW AGE

NEW AGE INTERNATIONAL (P) LIMITED, PUBLISHERS

New Delhi • Bangalore • Chennai • Cochin • Guwahati • Hyderabad
Jalandhar • Kolkata • Lucknow • Mumbai • Ranchi

NEW AGE INTERNATIONAL (P) LIMITED, PUBLISHERS
4835/24, Ansari Road, Daryaganj,
New Delhi - 110 002
Visit us at : www.newagepublishers.com

Offices at :
Bangalore, Chennai, Cochin, Guwahati, Hyderabad, Jalandhar,
Kolkata, Lucknow, Mumbai and Ranchi

Rs. 90.00

ISBN : 81-224-1585-7

Published by New Age International (P) Ltd.,
4835/24, Ansari Road, Daryaganj, New Delhi-110 002 and
printed in India at Ajit Printers, Delhi
typeset by Pagitek, Delhi

Dedicated
to
my parents
Late Sri. C. Venugopal
and
Late Smt. Sarojini Venugopal

Preface

This book Comprehensive Biotechnology–I is written as per the revised semesterwise syllabus of Biotechnology for the first year B.Sc. (Vocational) for semester students of Bangalore University. In this semester, the students of first year B.Sc. Biotechnology will have to answer one paper—Part A: Cell Biology and Part B: Genetics. This book covers the syllabus of Ist semester [illegible] Genetics.

This book contains upto date exhaustive information and written in a simpler manner that should enable the student to understand easily.

I have received a warm response [illegible] teachers. I hope that this book would [illegible]

I am thankful to M/s. New Age International [illegible] President, The Federation of [illegible] Director, New Age International (P) Ltd. Publishers [illegible]

Suggestions for further improvement of the book [illegible] acknowledged and incorporated.

Preface

This book Comprehensive Biotechnology–I is written as per the revised semesterwise syllabus of Biotechnology for the first year B.Sc. (Vocational) Ist semester students of Bangalore University. In this semester, the students of first year B.Sc. Biotechnology will have to answer one paper—Part A: Cell Biology and Part B: Genetics. This book covers the syllabus of Ist semester i.e., Cell Biology and Genetics.

This book contains upto date exhaustive information and written in a simpler manner that should enable the student to understand easily.

I have received a warm response for my contribution in Biotechnology–II book from students and teachers. I hope that this book would also generate a similar response from the students and teachers.

I am thankful to M/s New Age International (P) Ltd., Publishers, New Delhi, Mr. R.K. Gupta, President, The Federation of Educational Publishers in India and Mr. Saumya Gupta, Managing Director, New Age International (P) Ltd., Publishers, New Delhi for the cooperation.

Suggestions for further improvement of the book and correction of errors if any, would be gratefully acknowledged and incorporated.

Dr. V. Sreekrishna

Contents

PART A: CELL BIOLOGY

PART B

Contents

Part A: Cell Biology

1

Cells as a Basic Unit of Living System: Discovery and the Cell Theory

The cell can be understood by studying its complexity and functions. In the beginning the term itself was coined by the invention of the simple microscope by Anton Von Lewenhock and Jenssenis and the subsequent invention of the microscope by Robert Hooke in 1665; he introduced the term Cell while describing the microscopic texture of the cork.

The term Cell comes from the Latin word *Cellula*, which means small compartments. The improvement in microscopes and the techniques of staining helped us examining the structure and content of the cell.

The structural living unit other than the virus is the cell, which consists of smaller structures called Organelles, namely:

Cell Wall.

Plasma Membrane.

Cytosol, Cytoplasm

Golgi Bodies

Endoplasmic Reticulum,

Ribosomes.

Cytoskeletal structure.

Mitochondria.

Chloroplasts.

Lysosomes and Peroxisomes and

Nucleus.

Each organelle has a specialised function within the cell. A Cell containing these organelles and a well defined nucleus is called a Eukaryotic Cell.

The cells of bacteria and blue green algae lack a definite nucleus and organelles like eukaryotes, and are called prokaryotes. Many plants and animals exist as single cells; hence, the cell is the basic unit.

Three requirements have to be fulfilled for a cellular organisation. These are:

(i) a selectively permeable boundary separating the cellular components from the outside.

(ii) a metabolic system which would allow a series of enzyme catalysed reactions, and

(iii) a genetic information system which would store and transmit hereditary information for future generations.

However, viruses are exceptional as they possess a molecule containing the information required for reproduction and a capsular coat around this molecule. Viruses, however, cannot reproduce without a larger host which has an efficient biological machinery. For this reason they do not conform to a truly living system, and may be called semi living particles. Viruses have simple life histories. While the capsular coat is made up of protein, the information molecule is a nucleic acid DNA or RNA.

EUKARYOTES

They are highly evolved cells, possessing specialised organelles viz., endoplasmic reticulum; the endoplasmic reticulum may be granular or agranular. Ribosomes are deposited on the granular endoplasmic reticulum. Golgi complex, consisting of stacks of lamellae and vesicles of different shapes and sizes act as an intracellular pump to regulate fluid movement, expulsion of excretory products.

Lysosomes, phagosomes and pinocytosomes appear as irregular vesicular bodies. Lysosomes are rich in hydrolytic enzymes which carry on cellular digestion. Phagosomes contain semisolid or solid nutritive materials which are engulfed by the plasma membrane. Pinocytosomes are microscopic vesicles, pinched off from the plasma membrane like phagosomes; they too contain nutritive material but in liquid phase.

Mitochondria are most important organelles; they may be capsule or rodlike in shape. They are enveloped by a double membrane. The inner membrane, invaginates to form compartments. The mitochondria are responsible for cellular respiration. They also function as energy transducing organelles; contain enzymes of electron transport chain; sites for oxidation or carbohydrates, lipids and amino acids.

Plastids, which are characteristic of plant cell are another cell organelle; plastids without pigment are known as leukoplasts and those possessing colored pigments are known as chromoplasts. The plastids possessing green chlorophyll pigments are known as chloroplasts, which takes part in the photosynthetic activity of the cell. The leucoplasts serve for storage of starch, oil droplets and proteins.

The cytoplasm of the cell also contains vacuoles, which are large in plant cells but in animal cells they are small. The vacuoles are filled with liquid matter.

Another major component of eukaryotic cells is membrane bound neuclear complex. The nucleus consists of the nucleoplasm in which chromosomes and nucleolus are present. The nuclear membrane is double layered, with minute pores and nuclear membrane has direct contact with granular endoplasmic reticulum. Deoxyribonucleic acid (DNA) and the basic protein called histone are present in the chromosomes.

The Centriole is another organelle, that participates in cell division. They are spherical in shape located near the nucleus.

DISCOVERY

The term 'Cell' was coined by Robert Hook (1635-1703) while he was observing a thin layer of cork, which showed empty chambers lined by thick walls having an arrangement similar to a honey comb. He named these chambers 'Cells' (Cellula in latin means little rooms). Though he coined the term 'Cell' he could not study the internal aspects and the dynamic nature of the cell as he was observing dead/empty cells.

Later on a cell was found to be the fundamental unit of all living matter. A single cell is an entity, isolated from other cells by a cell membrane (and sometimes a cell wall) and contains a variety of chemical materials and sub-cellular structures Schleiden and schwann first described the cell and put forward some postulates, together which were called "The Cell Theory". According to this theory:

(a) Plants and animals are composed entirely of cells or substances produced by cells.
(b) All cells arise from pre-existing cells.
(c) The cells of which organisms are composed have their own life.
(d) The life of individual cells is subject to the life of the organism as a whole.

Later on, in the middle of 19th century, Purkinje (1840) coined the term "Protoplasm" for the substance inside the cell. In 1963 A.G. Loewy and P. Siekevitz defined cell as "a unit of biological activity delimited by a semipermeable membrane and capable of self reproduction in a medium free of other living systems." Cells have various functions. They contain protoplasm which is majorly composed of water, protein and carbohydrates and also minute amounts of fats and storage compounds. They also contain Nucleic acids which carry information from one generation to the other. All these molecules are collectively called "Macromolecules". Macromolecules are present in almost all cells and hence we can assume that cells have orginated from a common origin i.e. the ancestral cell.

An entity of protoplasm bound by a barrier that separates the interior from the surrounding may be called a "Cell"; but in the true sense a "Cell" should posses certain unique characteristics.

1. A cell should be able to nourish itself from the surroundings i.e. self Nourishment.
2. It should be able to use the nutrients and grow in size and multiply i.e. growth
3. A cell should be able to differentiate i.e. mature and perform its functions efficiently which involves structural and physiological changes, i.e. differentiation.
4. A cell should be able to respond to stimulus. It should also be able to cooperate and coordinate with other cells to perform its functions i.e., it should express the phenomenon of cell signalling.
5. Lastly a cell should evolve i.e. change for the better.

These characters are the Hallmarks or unique characters of a cell.

A cell is a dynamic system. The cytoplasm of the cell contains many substances which are actively involved in the metabolism of the cell. The cytoplasm is always active as various chemical reactions, biosynthetic reactions and other mechanism like transport etc. are always operating. The composition of the constituents of the cell is unique and specific. This composition differs from the surrounding environment. There is constant pressure upon the cell and its constituents by the environment. The cell wall or the cytoplasmic membrane give protection to the cell. The cell tends to balance the pressure and tend to be in equilibrium with the environment. The cell takes in nutrients and other substances from the surroundings and secretes out waste products and other solutes from the interior. So it is not a closed system. A cell thus is a dynamic open system; which has a specific chemical composition and tends to be in equilibrium with the surroundings. Though the cell tends be in equilibrium, it is, in the true sense a Non-equilibrium system as it has many different biological activities occuring inside it and also is able to replicate on its own, protected from the environment by a membrane and exists as a non-equilibrium system.

Basing upon their structural physiological and evolutionary progress cells are classified into Eukaryotes and Prokaryotes. Prokaryotes are primitive cells which have simple structure and organisation. Bacterial and Blue-green Algae (Cyanobacteria) are included in this group. Eukaryotes on the other hand are well organised complex cells which are highly evolved. Fungi, Algae and higher organisms (plants, animals) are included into this group. Viruses can neither be included into prokaryotes nor into Eukaryotes as they are not cells at all. They lack many attributes of cells, of which the most important is that they are not dynamic open systems. A single virus particle is a static structure, quite stable and unable to change or replace its parts. Only when it is associated with a cell does a virus become able to replicate and acquire some of the attributes of a living system. Thus, unlike cells viruses have no metabolism of their own. Cells have been defined into three evolutionarily distinct cellular lineages, the Bacteria, Archea and Eukarya. Bacteria and Archeae (primitive bacteria) are prokaryotic cells and Eukarya and Eukaryotic cells.

POINTS TO REMEMBER

1. A cell is the fundamental unit of all living organisms.
2. 'Robert Hooke' coined the term "cell", and "protoplasm" was coined by Purkinje.
3. Cells are dynamic open systems.
4. Cells should be able to Nourish themselves grow differentiate, co-operate and co-ordinate with other cells and most importantly evolve.
5. Cells tend to attain equilibrium but due to the structural and biological compositions they remain as non-equilibrium systems.
6. These cells basing upon their evolutionary progress are categorised into Bacteria, Archeae and Eukarya. The prokaryotes are Bacteria, Cyanobacteria and Archeae (Primitive Bacteria) and Eukaryotes include Fungi, Algae and Higher Organisms.

THE CELL THEORY

In 1665 Robert Hooke suggested the term "cell" while describing the microscopic texture of the cork. This was followed by illustrations conforming to cellular nature of living organisms by a number of workers during the next two centuries.

Schleiden and Schwann are generally accredited as the exponents of the "cell theory", which states that both plants and animals are made up of aggregations of basic units called cells.

Schleiden and Schwann writing in the 19th century, introduced the cell theory, which can be summarised as follows:

(i) Organisational cells are differentiated into distinct cell types.
(ii) Nucleus is a common feature of all cells with the exception of viruses and red blood cells.
(iii) The protoplasm is the living content of the cell, and determines the activity of the cell and of the whole organism.
(iv) New cells originate from pre-existing cells through division.

Life can be considered as a unique and complex interaction of matter and energy that assumes a hierarchy of structural order. The biological continuity, thus, consists of different levels of biological organisation. Atoms, the smallest units of matter, interact to form molecules that may be variously joined to form more elaborate macromolecules characteristic of living organisms. The interacting macromolecules often result into structures called organelles and the next higher level of organisation is the cell. The cell is the smallest unit of interacting matter and energy that exhibits properties of life.

However, the chemical components of the living cells are qualitatively quite different from those of the physical environment. Majority of the molecules in living cells are organic compounds, which occur in great variety and complexity.

For example in *Escherichia coli* bacterium (Fig. 1.1), there are more than 5, 000 types of organic molecules, of which 3,000 are proteins and 1,000 are nucleic acids. *Homo sapiens,* i.e., Man is known to have more than 1,00,000 types of proteins. This immense variety of organic compounds is reducible to a surprising simplicity; all proteins are composed of 20 types of linearly arranged amino acids. Similarly all nucleic acid molecules are composed of four types of nucleotides.

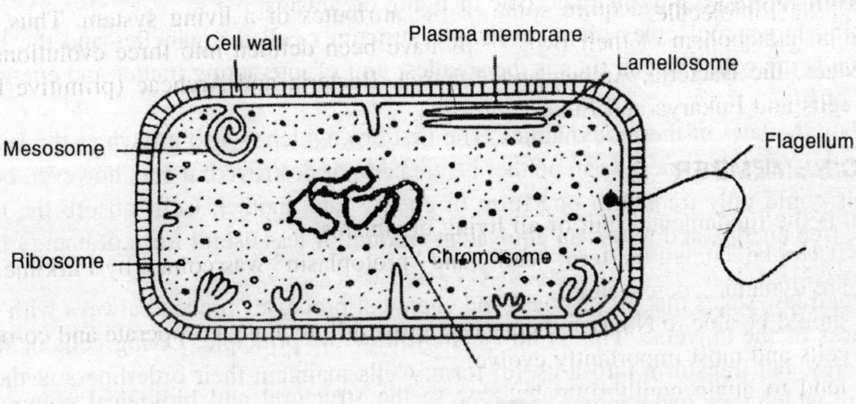

Fig. 1.1. A typical Escherichia coli cell showing prokaryotic organisation

These building blocks, thus, serve as the raw material for macro-molecular synthesis and perform various functions. Amino acids, for example, also serve as the raw material for macromolecular synthesis and perform various functions.

Further, Amino acids also serve as precursors of hormones, alkaloids, pigments, etc. Strange as it may appear, the building blocks are identical in all living organisms. Therefore, the following generalisations can be formulated:

(A) There is an underlying simplicity in the molecular organisation of all living cells.

(B) There is an underlying principle of molecular economy in living cells.

(C) All living organisms have a common ancestor.

PARAMETERS OF LIFE

Though it is not possible to define life, a set of properties can be suggested which characterise life.

(a) *Self Organisation:* Living systems are self organising, whereby specialised structures would have metabolic cooperation in an organised manner.

(b) *Self Regulation Capacity:* Self regulation of metabolic processes should be expressed by living systems. Proteins, which also function as enzymes, are capable of performing polyeric synthesis and degradation

(c) *Mechanism of Energy Transfer:* Living systems obey laws of thermo-dynamics. They have the capacity to liberate energy which is released through degradation of food materials, and are conserved in the form of ATP. Energy of these phosphate groups is utilised for different metabolic activities.

(d) *Self Reproducing Capacity:* All living organisms reproduce their own kind during their limited lifespan. DNA molecule helps in this process and also in passing genetic information to daughter cells.

(e) *Isothermal Open System:* Living organisms are capable of exchanging matter and energy within the environment; chemical transformations of matter occur at a temperature of around 37°C in majority.

(f) *Adaptation Capacity:* Genetically, living systems are capable of adapting to new environments or improving their efficiency within the same environment and causing evolutionary variations.

Life could be considered as a unique and complex interaction of matter and energy which possesses a sequence of structural order. The biological continuation consists of various levels of biological organisation. Atoms, the smallest units of matter interact to form molecules that may join to form more complicated macromolecules characteristic of living organisms.

Interacting larger molecules often result into structures called organelles and the higher level of organisation is the cell. The cell thus is the smallest unit of interacting matter and energy that exhibits properties of life.

Cells obey the laws of thermodynamics. The first law, which is also known as the law of conservation of energy, states that energy can neither be created nor destroyed; it can, however, be transformed. Living cells could only transform one form of energy into another. Cells absorb the useful form of energy, i.e. free energy, and return an equivalent amount of less useful form of energy to the environment.

The second law states that all physical and chemical processes proceed always with an increase in disorderliness of the universe. This is how cells follow the principle. Living cells neither create nor destroy energy, but transform into a useful form. Cells maintain their orderliness at the cost of their environment which they make more disorderly.

SUMMARY

It is believed that the ancestral organisms both plants and animals gained their energy from chemotrophs and later specialised in synthesising own food through photosynthesis.

Structural life unit other than the virus is the cell. A cell consists of smallest structures called organelles, each having a specialised function within the cell. A cell consisting of cellular organelles and well defined nucleus is called eukaryotic cell.

The cell of blue green algae, and bacteria take a definite nucleus and organelles like eukaryotes; they are thus called prokaryotes. As many plants and animals exist as unicellular organisms.

Several cell theories proposed are discussed.

EXERCISE:

1. Why we should not consider virus as a living entity?
2. What are the three main requirements for a cellular organisation?
3. Explain in detail the cell theory?
4. Write shortnotes on:
   (a) Prokaryotes
   (b) Eukaryotes
   (c) Parameters of life.

2

Surface Architecture

PLASMA MEMBRANE

The structure that separates the cell content from the external environment is the plasma membrane. It is a thin film (6 to 10 nm thick) constituting continuous lipid bilayer with proteins intercalated in or adherent to both surfaces. The plasma membrane can be resolved only with the electron microscope, which reveals its numerous infoldings and differentiations as well as the different functions that establish connections with neighbouring cells. The main function of plasma membrane is to control selectively the entrance and exit of the materials. The plasma membrane is covered and reinforced by the cell wall in plant cells and by the cell coat in animal cells.

The cell wall is responsible for the rigidity of most plant tissues. This structure mainly consists of cellulose fibres into which other substances like lignin, which is the main component of timber may be incorporated. There are funnels ranging through the cell wall that is plasmodesmata which allow communication with other cells in a tissue. The cell coat covered in most animal cells is made of glycoproteins, glycolipids and polysaccharides that may extend the thickness of the cell membrane and continue far beyond it. Besides protection, the cell coat is involved in molecular recognition between cells; it contains enzymes and antigens and is fundamental in the association of cell in a tissue.

Cell membrane term was coined by Nageli and Cramer in 1865 and Plowe coined the term plasmalemma to plasma membrane.

In 1902 Overton postulated that the plasma membrane is composed of a thin layer of lipid. In 1926 Gorter and Grendel found that the lipid content of hemolyzed erythrocytes was sufficient to form a double layer of lipid molecules over the entire cell surface.

Other information comes from the study of interfacial tension of different cells. Tension at a water-oil interface is about 10-15 dyn/cm, whereas surface tension of cells is almost nil. The low tension is due to the presence of protein among the lipid components. In fact, when a very small amount of protein is added to a model lipid water system, the surface tension is lowered.

To explain all these, many molecular models of cell membrane were proposed by many cell biologists. Some of them are:

(a) *Lamellar model:* Danielle and Dawson proposed that the plasma membrane is sheath like and composed of two layers of lipid which are sandwiched between two continuous layers of proteins. This is known as sandwich model or Trilamellar model (1935).

(b) *Unit membrane model:* Robertson (1959) proposed his famous unit membrane model. According to him, the unit membrane consists of bimolecular lipid leaflet sandwiched between outer and inner layers of proteins organised into the pleated sheet configuration (protein-phospholipid-phospholipid-protein). Such an arrangement was presumed to be basically the same in all cell membranes. The observations of thin section by electron microscopy has led to the concept of trilayered plasma membrane which is also called the unit membrane model.

c) *Micellar model:* According to this model a central core (micelle) of hydrophobic phospholipids is surrounded by the hydrophilic protein molecules.

d) *Protein crystal model:* According to this model the phospholipids are intercalated in the gaps or cavities of crystal like protein structure.

All these models and unit membrane model, however, are certainly an oversimplification.

Gorter and Grendel in 1934 performed an experiment to study the arrangement of the components of the plasma membrane.

A known number of RBCs were taken and lysed by treating the RBC with hypotonic solutions. This was done in beakers containing water. After the lysis of the cells the air-water interphase shows a layer of phospholipid molecules spread over the surface of water. As they already knew the number of RBC they could approximately measure the area occupied by the membranes of the RBC on the air-water interphase which now had a single layer of phospholipid molecules. However, they found that the area occupied by the single layer of phospholipid molecules on the air-Water interphase was double the area occupied by the membranes in RBC as per calculation. This led to the conclusion that the phospholipids were not present in a single layer but a double layer.

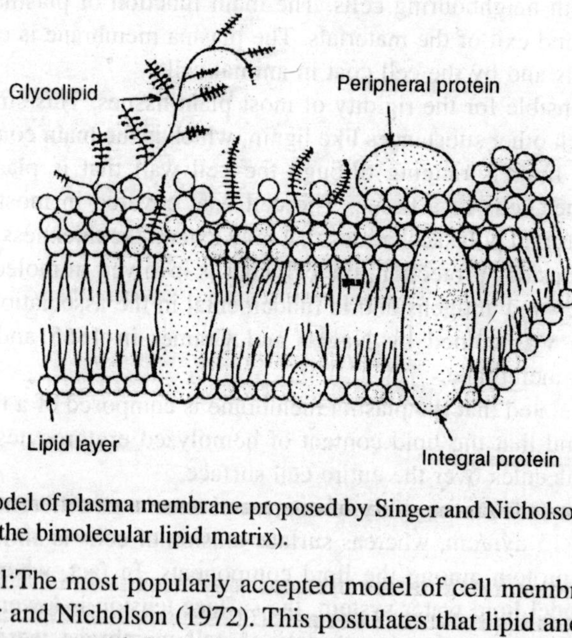

Fig. 2.1. Fluid-mosaic model of plasma membrane proposed by Singer and Nicholson (globular proteins are seen floating in the bimolecular lipid matrix).

The fluid mosaic model:The most popularly accepted model of cell membrane is the fluid mosaic model proposed by Singer and Nicholson (1972). This postulates that lipid and proteins are dispersed in a kind of mosaic arrangement and that biological membranes are quasi-fluid structures in which both lipids and the proteins are able to move within the bilayer, i.e, the proteins are embedded in a lipid bilayer in such a way that proteins float in lipid sea. (Fig. 2.1). The lipid molecules mainly consist of phospholipids which are arranged in two layers. Each phospholipid molecule is made up of a glycerol head connected to a tail of long chain fatty acid-hydrocarbons, which always contain 1 or 2 double bonds and give the tail rather a kinked structure.

Many types of proteins, namely extrinsic, intrinsic and transmembrane proteins are present interpersed with the lipid molecules. Extrinsic proteins are the proteins which are superficially attached to the lipid molecules. The integral proteins of the membrane are intercalated to a greater extent into a rather continuous lipid bilayer. These integral proteins are amphipathic with polar regions (hydrophilic) protruding from the surface and nonpolar (hydrophobic) regions embedded in the hydrophobic interior of membrane (Fig. 2.2).

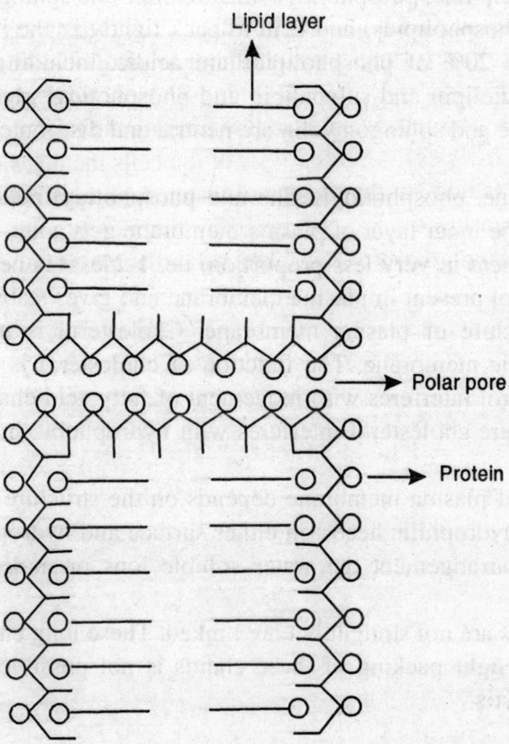

Fig. 2.2. Danielli-Davson model of plasma membrane showing pores in the membrane

Proteins and Phospholipids are not tightly associated with each other and have freedom to move along the surface of plasma membrane. Phospholipid molecules may also move along the cell membrane thus making the plasma membrane a viscous fluid rather than a rigid structure. In fact, the cohesive forces between lipids and proteins are relatively weak interactions, i.e. ionic hydrogen bonds and mainly hydrophobic in character. Some of the lipids and proteins on the outer surface are frequently associated with sugars to form glycolipids and glycoproteins.

COMPOSITION OF PLASMA MEMBRANE

In general, the plasma membrane is composed of

(i) lipids (Phospholipids, cholesterol)

(ii) proteins

(iii) oligosaccharides bound to lipids (glycolipids) and proteins (glyco-proteins).

Lipids: The main lipid components of the plasma membrane are phospholipid, cholesterol and glycolipids. Their proportions vary in different cell membranes.

Phospholipids: They are loosely aranged and form 20-79% of the cell membrane. The major proportion (50%) of membrane phospholipid is represented by

(a) Phosphotidylcholine-neutral
(b) Phosphotidyl ethanolamine-neutral
(c) Sphingomyelin-neutral
(d) Phosphatidyl Serine- Acidic, -ve charge

Among these phosphotylcholine, phosphotidyl ethanolamine and sphingo-myelin have no net charge at neutral pH (i.e, neutral phospholipids) and tend to pack tightly in the bilayer. This property is also shared by cholesterol. 5 to 20% of phospholipids are acidic, including phosphotidyl inositol and phosphotidylserine and cardiolipin and sulphalipin and phosphotidyl glycerol. They have a negative charge. Phosphotidyl choline and sphingomyelin are neutral and distributed more on the outer layer of plasma membrane.

Phosphotidyl ethanolamine, phosphotidylserine and phosphotidyl inositol are distributed more in the inner side ; due to this the inner layer of plasma membrane gets a net -ve charge on it. Though the phosphotidyl inositol is present in very less proportion, i.e. 1-2%, it is helpful in cell signalling.

Cholesterol: It is the sterol present in plasma membrane and gives stability or acts as strengthening agent to preserve the structure of plasma membrane. Cholesterol is present in equimolar to the phospholipids in cytoplasmic membrane. The function of cholesterol is temperature dependent . At higher temperature cholesterol interferes with movement of fatty acid chain and prevent permeability of solutes. At low temperature cholesterol interferes with hydrophobic interactions avoiding freezing of the membrane.

Basic functional ability of plasma membrane depends on the structure of phospolipid molecules:

(i) Phospholipids have hydrophilic heads on either surface and hydrophilic tails are present in the centre. Due to this arrangement the water soluble ions or molecules cannot enter through plasma membrane.
(ii) The hydrophobic tails are not straight but are kinked. These long chains of fatty acids can move freely due to which tight packing of these chains is not possible. Phospholipids are able to rotate on their own axis.

MEMBRANE PROTEINS

Proteins represent the main component of most biological membranes. Numerous enzymes, antigens and various kinds of receptor molecules are present in plasma membrane. These proteins were identified by Singer and Nicholson. The membrane may consist of 30-50% of proteins. For every 50-100 molecules of lipid one protein molecule is present. Membrane proteins have been divided into peripheral and integral proteins according to the degree of their association with the membrane, and the methods by which they can be solubilized.

PERIPHERAL PROTEINS

They are present only on the surface of the plasma membrane, due to their hydrophilic nature, and are bound to the plasma membrane by ionic bonds and protein-protein interactions. They are easily dissociated by mild treatments like high pH, hypotonic solutions, high ionic strength, metal ion chelating agents, etc., The dissociated proteins are soluble in water and are usually free of lipids and molecularly dispersed in neutral aqueous buffers. Their dissociation will not cause actual disruption of the plasma membrane; for example, i) spectrin may be removed from red cell ghosts by chelating agents. ii) cytochrome-C found in mitochondria is easily removed in high salt solutions.

INTEGRAL PROTEINS

They represent more than 70% of the membrane proteins. They pass through the membrane and are associated with hydrophobic regions of the lipid bilayer also. They are exposed to either side of the membrane, and when present throughout the membrane they are called transmembrane proteins and consist of both hydrophilic and hydrophobic regions, i.e. they are amphipathic. They are dissociated from the plasma membrane by surface acting agents like detergents, hydrophobic bond breaking organic solvents and chatropic agents. Their dissociation is accompanied by the dissociation of the whole plasma membrane.

OTHER PROTEINS

The oligosaccharide group of glycoprotein formed from the ribosomes attached to endoplasmic reticulum integrates with phosphotidyl inositol present in the inner leaflet membrane by GPI anchor (Glycosyl Phosphotidyl Inositol Link) to form a GPI molecule. These are known as GPI linked proteins. Other than these, some more proteins are seen in plasma membrane formed or synthesized by free ribosomes. In erythrocyte or leucocyte membrane, the leucoproteins consists of Hexose, Hexosamine, Fucose, NAG and NANA-N-acetyl-neuraminic acid (sialic acid). The proteins which contain sialic acid in their glyco group are called sialo proteins and these proteins are recognised by endothelial receptors. Because of the amphipathic nature of the transmembrane proteins, they cannot be easily crystallised and hence their 3 dimensional structure cannot be identified. The transmembrane protein has 3 regions: The middle portion is α-helical structure; above and below it, peripheral proteins are present which accept light and send to L,M,N portions of α-helical structure. It consists of single polypeptide chain of 247 amino acids, which traverse 7 times the lipid bilayer. Each of the traversing segments is an α-helix with some 25 amino acid residues. The oligosaccharides of the proteins and lipids on the outer side of the membrane form a layer around the membrane called Glycocalyx in eukaryotes, and its basic function is to protect and also work as binding sites or receptors.

FUNCTIONS OF THE PLASMA MEMBRANE

Main functions of plasma membrane are:

(i) Cell Permeability

(ii) Proteins present on cell membrane may act as carriers for transport of materials and may help in cell recognition or may act as receptors for informational molecules.

CELL PERMEABILITY

Permeability is fundamental to the functioning of the living cell and to the maintenance of satisfactory intracellular physiological conditions. This function determines which substances can enter the cell, many of which may be necessary to maintain its vital processes and the synthesis of living substances. It also regulates the outflow of excretory substances and wastes from the cell.

The presence of membrane establishes a net difference between the intracellular fluid and extracellular fluid. This may be fresh water or salt water in unicellular organisms growing in ponds or seas, but in multicellular organism, the internal fluid i.e. blood, lymph and especially the interstitial fluid, is in contact with the outer surface of the cell membrane. One of the functions of the cell membrane is to maintain a balance between the osmotic pressure of the intracellular fluid and that of the interstitial fluid.

Passive transport: Passive transport occurs along the lipid membrane depending upon the concentration gradient and does not involve any proteins of energy. The molecules move across the lipid membrane from higher to lower concentration. Molecules like water can pass through passive transport. Small polar molecules and hydrophobic molecules are permeable by passive transport whereas the charged molecules, larger molecules and certain hydrophilic molecules cannot pass through passive transport.

Molecules like oxygen, carbon-di-oxide and polar molecules like water, nonpolar solvents like Benzene and alcohol are permitted. Amino acids are not permitted as they are large and charged. Glucose is also not allowed due to its larger size. Small ions like H^+, Ca^{+2}, Cl^-, Na^+, K^+ are also not allowed by passive transport due to their charge (Fig. 2.3).

FACILITATED TRANSPORT (DIFFUSION)

Occurs by two different ways—

(a) occurs along the concentration gradient.

(b) occurs against the concentration gradient.

Proteins are involved in this transport. Two types of protein transporters are seen.

(i) carrier proteins

(ii) channel proteins

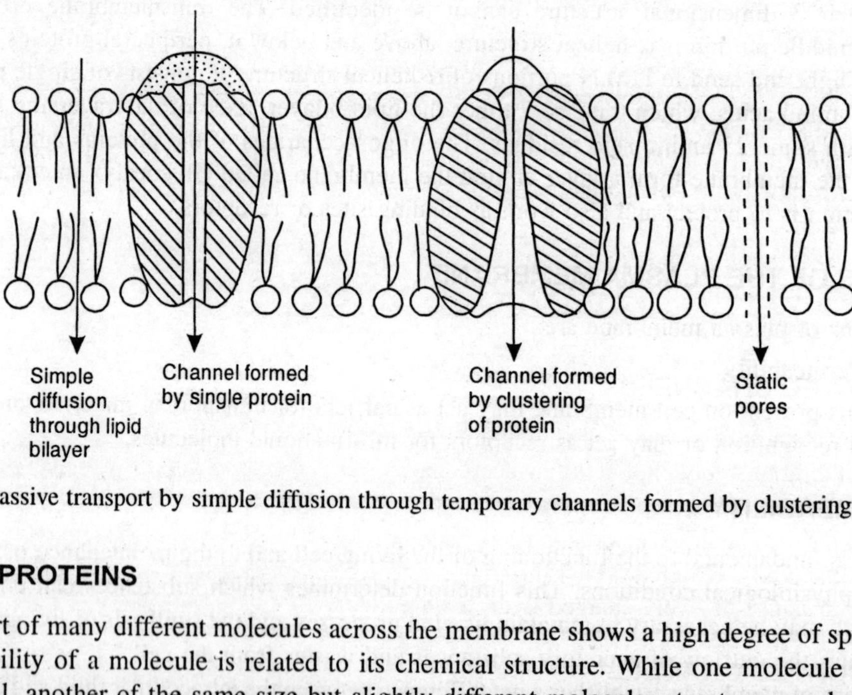

Fig. 2.3. Passive transport by simple diffusion through temporary channels formed by clustering of proteins

CARRIER PROTEINS

The transport of many different molecules across the membrane shows a high degree of specificity, i.e, the permeability of a molecule is related to its chemical structure. While one molecule may readily enter the cell, another of the same size but slightly different molecular structure may be completely avoided. A clear example of the selectivity is provided by the transport of two isomers (Glucose and Galactose) into a bacterial cell. Although the only difference between them is in the position of -OH group at carbon 4, these two sugars penetrate the membrane by different transport mechanisms. This type of selectivity is attributed to transport proteins or carrier proteins or permeases. It is thought that a permease functions in a manner somewhat similar to an enzyme or a receptor, in having a binding site able to recognize the molecule to be transported. Some permeases can transport only if there is a favourable concentration gradient. Glucose concentration is always less inside the cell, as it is actively metabolised inside the cell, except in liver where active outward flow of glucose molecules is seen.

The carrier mechanism has also been postulated to mediate the entrance of the glucose into intestinal cells. This mechanism is activated by Na^+. The carrier ion in this model has polar amino acids or binding site for Glucose, and should exist in two conformational states, one with low affinity for glucose and another with high affinity. A carrier model can be postulated in which the glucose site is initially with a high affinity state and binds the glucose molecules at the surface of the cell. In the second step the carrier is translocated and the glucose site changes to a low affinity releasing the molecule into the cell. The two affinities are modulated by Na^+ concentration inside and outside the cell. This carrier mechanism follows fixed pole mechanism in which the carrier is represented by integral proteins that traverse the membrane, and which once bound to the molecule to be transported undergo conformational changes.

CHANNEL PROTEINS

Special channels meant for the transport of water are present in the membrane. Ex.Aquaporins. Porins are present in the outer membrane of Eukaryotes. The transport through these channel proteins is much faster than the passive diffusion.

ION CHANNELS

Ion channels form the most important means of transport of various ions and water. They are more effective and efficient due to the following reasons:

(1) Faster rate of movements of ions i.e, up to a million ions can pass through it per second.
(2) They are specific to change and size of ions.
(3) They are not always open.

Based on the type of signals they require to close and open, they are of two types:

(i) Ligand gated channels
(ii) Voltage gated channels

LIGAND GATED CHANNELS

They open and close in response to the binding of neurotransmitters or some other signaling molecules.

Ex. The binding of Acetylcholine to the acetylcholine transport molecule determines the opening or closing of the negatively charged hydrophobic gates. Hence acetylcholine acts as signalling molecule and when it binds to the molecule, the gates open due to conformational change and allow only +vely charged ions like Ca^{+2}, K^+, etc. to pass through them. But –vely charged ions are not allowed to pass through them because of –vely charged hydrophobic gates.

VOLTAGE GATED CHANNEL

The opening and closing of the voltage gated channel depends upon the electron potential or charge of the cell Ex. Na^+ pump, K^+ pump.

NA$^+$ PUMP

Generally Na^+ are pumped outside the cell and the Na^+ pump aids in this.

Generally Na^+ size is small and appropriate to the size of the Na^+ pump and so passes through it along with water. But as K^+ is larger than Na^+, K^+ cannot pass through a Na^+ pump.

K^+ PUMP:

Generally K^+ are pumped inside the cell and the channel of K^+ pump has a binding site and so it can pass through it along with water. But Na^+ size is very small to bind with the active site in K^+ Pump, and hence only K^+ions can pass through them.

GROUPS TRANSLOCATION:

Translocation of a molecule inside the cell, which involves the chemical alteration of that molecule,is called Group translocation. Group translocation always occurs against the concentration gradient but without true expenditure of energy. Ex. Glucose is transported by group translocation into cell as glucose-6 phosphate by addition of a phosphate molecule outside the cell. Here the ATP is used up in the phosphorylation of Glucose, i.e. for biochemical reaction but not actually for transport. This glucose 6 phosphate may follow Glycolysis in the cell. Sugars are generally transported by this method.

ACTIVE TRANSPORT

Active transport involves transport of molecules via the cell membrane with expenditure of energy; energy procured from ATP hydrolysis or electron potential of the cell, without the chemical alteration of the molecule against the concentration gradient. Ex. Na^+, K^+ pump. It includes protein molecules with 3 binding sites for Na^+ and 2 binding sites for K^+ and can pump 3 Na^+ outside and $2K^+$ inside the cell by utilisation of only one ATP molecule.

In normal configuration the Na^+ binding sites are exposed to the cytoplasm and the Na^+ which are always pumped outside the cell bind to these binding sites when the protein molecule is phosphorylated by the hydrolysis of ATP. As soon as the Na^+ binds to the active sites, conformational change takes place and the Na^+ binding sites are exposed to the cytoplasm and so K^+ from outside the cell binds to these active sites which causes dephosphorylation of protein molecule and thus leading to conformational change. The K^+ are released inside the cell. So only the ATP molecule is utilised to pump out 3 Na^+ and to pump in 2 K^+ ions by active transport.

CA2$^+$ PUMP

Ca2$^+$ Pump is also similar in structure to the Na^+, K^+ pump and helps in active transport of Ca2$^+$ ions outside the cell by ATP hydrolysis. The concentration of Ca^{+2} inside the cell is 0.1 mm. The concentration of Ca^{+2}outside the cell = 1mm. But the Ca^{2+} are transported outside the cell. During this transport some other ions may be transported inside the cell simultaneously.

CYSTIC FIBROSIS

It is a disease caused by malfunctioning of ion channels namely CFTR i.e. Cystic Fibrosis Transport Regulator Channel.

The cells are very sensitive to the Ca^{+2} concentration. Ca^{+2} acts as a

- cell signalling molecule.
- H Pump is more frequent in plant cells and bacteria.

– Na^+, K^+ pumps are frequent in animal cells.
– The entry and exit of the water molecules in the cell is governed by the turgor pressure which develops due to the presence of a rigid cell wall.

ION GRADIENT ACTIVE TRANSPORT

The concentration of ions inside or outside the cell is called ionic gradient. The cell always maintains the ionic gradient by pumping of ions like Cl^-, Ca^{+2}, Na^+, H^+, K^+. If +ve charge is formed outside of the cell, OH^- ions are always present inside the cell having –ve charge. This is the stable gradient always present around the cell. Any cell maintains an ion gradient across its membrane by active transport. Cells contain a net +ve ion gradient on the outer surface, where large amount of H+ and +ve charged ions are accumulated. The inner surface of the cell membrane maintain a -ve gradient by the accumulation of OH- ions. The ion gradient keeps fluctuating across the membrane due to the movements of ions into or outside the cell against the concentration gradient by active transport. When Na^+ and Ca^{+2} is transported out of the cell, against the concentration gradient, there is a reduction in net ionic gradient, across the cell membrane. The movement of cation across the cell membrane is thus compensated by the inward flow of equal number of cations. This mechanism of maintaining ionic equilibrium in the cell membrane is called "Donnan's equilibrium.

TRANSPORTERS

Three types of transporters are seen in the cell membrane namely:

(a) Uniporter
(b) Symporter and
(c) Antiporter

Uniporter: This type of transporter transfers only one specific molecule either outside or inside the cell through it. Ex. Na^+ Pump.

Symporter: This type of transporter transports one type of ions and some other ion or molecule can also pass along these ions in the same direction. Ex. Glucose enters into the cell along with Na+ ions through Na+ Channel along the concentration gradient.

Antiporter: This type of transporter allows one type of ions inside and other type of ions outside the cell simultaneously. Ex. N^{a+}, K^+ Pump - Na^+ outside K^+ inside K+ - Ca^{+2} Pump-K+ inside Ca^{+2} outside the cell.

OTHER FUNCTIONS OF PLASMA MEMBRANE

Endocytosis: Endocytosis is related to the activity of plasma membrane. It may not take place in all the cells. It includes phagocytosis and pinocytosis.

Phagocytosis: The ingestion of the food particles inside the cell by the formation of a phagosome of vesicle by specialised cells called phagocytes is called phagocytosis, whereas Macrophages and neutrophils are called pinocytosis.

Process: Once the phagosome or pinosome are formed, they are associated with lysosome to form phogolysosome and pinolysosome into which the hydrolytic enzymes are released. Active digestion takes place and simpler compounds or waste materials are formed in the phagolysosome. The useful material is secreted in the cell and finally the phagosome with debris or waste material is sent or vomited out of the cell by a process known as exocytosis, emiocytosis (or) cell vomiting.

Vesicles: Vesicles are globular structures that bud off from the cell membrane during endocytosis. They are of two types: 1) Nonspecific Vesicles 2) Specific Vesicles.

1) *Nonspecific vesicles:* These are more or less equal to phagosomes or pinosomes, which may later combine with lysosomes, endoplasmic reticulum or golgi bodies depending upon the type of molecules or on the nature of the residue in vesicles.

- If it contains complex materials – it associates with lysosomes for digestion into simpler compounds.
- If it contains proteins – they are passed to endoplasmic reticulum for sorting, marking or for Glycosylation. If the proteins present are useful to the cell, then they are targeted to specific spaces; otherwise, they are thrown out if they are not required.
- chances of union of vesicles with golgi bodies is rare.
- Sometimes these vesicles may also combine with endosome (which is a vesicle that is present in the cell for a long time before forming secretory vesicles).

2) **Specific vesicles:** Specific vesicles contain receptors which facilitate them to capture particular material. Ex. Intake of cholesterol. There are certain pits in the mammalian cells called "clatherine coated pits", coated by a substance called clatherine, which contains LDL receptors. Cholesterol is ingested in the form of LDL – Low Density Lipoprotein. The LDL gets bound to the LDL receptors and slowly the pit will be filled with LDL. Then it forms clatherine coated vesicles which will be transferred specifically to the lumen of endoplasmic reticulum to get degraded to form cholesterol. This cholesterol is then passed to a specific site via endoplasmic reticulum, Golgi apparatus. Sometimes LDL may be sent outside the cell or may be loaded on the inner walls of blood vessels, thus reducing the width of the blood vessels and causing high blood pressure (B.P).

SPECIALISED STRUCTURES OF THE CELL MEMBRANE

The differentiations of the cell membrane corresponds to regions specially adapted to different functions such as absorption, secretion, fluid transport, mechanical attachment or interactions with neighbouring cells.

(i) **Microvilli:** In the intestinal epithelium microvilli are prominent and form a compact structure that appears under the light microscope as a striated border. They represent the cytoplasmic processes covered by plasma membrane or the projections of the plasma membrane. They are mainly involved in the absorption of nutrients and are supported by cytoskeletal elements like microtubules and microfilaments. The actin filaments are attached to the tip of the microvilli by -actinin and their function is to produce contraction in the microvilli. The outer surface of microvilli is covered by a coat of filamentous material (fuzzy coat) composed of glycoprotein molecules.

(ii) **Interdigitation:** These are structures formed by folding of the plasma membrane into finger like or digit like structures which are predestined to form other specialised structures.

(iii) **Special invaginations:** Special invaginations of plasma membrane are seen forming the chambers between them, and mitochondria are present in these chambers helping in the production of energy for transport.

(iv) **Zonula occludens** (tight junctions): Tight junctions are specially differentiated regions that seal the intercellular space thus preventing the passage of fluid to and from the lumen. These junctions are present just below the apical border of the cell. In a polarised epithelium, one of the functions of the tight or occluding junctions is the maintenance of different intercellular environments in the apex and basolateral regions. The tight junction could act as a barrier to the diffusions of macromolecules or lipids in the bilayers, so as to differentiate the apical from basolateral portion, of the cell membrane, each having their own particular composition and physiological characteristics.

Mechanical adhesion between cells is mainly done by the desmosomes, of which there are two types: belt and spot desmosomes. The difference between them resides in their localisation, extension and relationship with the microfilaments.

Belt desmosomes (Zonula adherence or Terminal bars): Intermediary junctions are generally found at the interphase between columnar cells, just below the region of tight junctions. They form a band that girdles the inner surface of the cell membrane. This band contains a web of (6nm) actin filaments and another group of interwoven, intermediate filaments of 10nm. Actin microfilaments are contractile and intermediate filaments play a structural role. The plasma membranes of the adjacent cells come closer to each other and cytoplasm gets thickened into a gel in this region.

Macula adherence (Spot desmosome): They appear as darkly stained bodies on the cell surface under the light microscope. They represent localised circular areas of contact, about 0.5 nm in diameter, in which the plasma membranes of the two adjacent cells are separated by a distance of 30 to 50 nm. The electron microscope reveals a midline structure or intercellular core and two parallel desmosomal/plasma membranes. Under each plasma membrane there is a discoidal intracellular plaque, towards which numerous filaments called tonofibrils or tonofilaments converge. These tonofibrils are not contractile and belong to common type of intermediate keratin filaments. They are rigid structures.

Hemidesmosomes: Along the basal surface of some epithelial cells separate desmosomes called hemidesmosomes may be observed. These are similar to desmosomes in fine structure, but represent only half of them, the outer surface frequently being substituted with collagen fibrils. Some tonofibrils may be present.

Gap junction: The Gap junction or nexus or communication junction appears as a plaque-like contact in which the plasma membrane of adjacent cells are in close-apposition or tightly connected, separated by a space of only 2 to 4 nm. The cytoplasm is absent between plasma membranes in this region. It mainly binds the cells together and helps in cell communication.

Inter membrane canaliculus or intercellular canaliculus: They are hexagonal structures formed by the infolding of the plasma membrane cytoplasm which is thickened at this region. It basically helps in maintaining communication between cells in the form of ions.

Cell wall: The cell wall constitutes a kind of exoskeleton that provides protection and mechanical support for the plant cell. This includes maintenance of a balance between the osmotic pressure of the intracellular fluid and the tendency of water to penetrate the cell. The cell wall of fungi is made up of chitin, a polymer of N-acetyl glucosamine. The exoskeleton of crustaceans (Ex. scorpion) and Insecta (Ex. cockroach) is also made up of chitin.

Cell wall of plants: The wall consists of a microfibrillar network lying in a gel like matrix of interlinked molecules. The microfibrils are mostly cellulose, consisting of straight polysaccharide. These are the glucose chains, which by intra, and intermolecular hydrogen bonding produce the structural unit known as microfibril. Microfibrils in turn bind among themselves with non cellulose polysaccharides and proteins to form the cell wall. In addition to protein, the gel matrix contains some polysaccharides and lignin. The major polysaccharide fractions are pectin substances (containing Galactose, Galacturomic acid, arabinose, sometimes Rhamnose may be present) which are soluble in water and Hemicellulose (consisting of glucoronic acid, xylose, fucose, mannose, glucose, etc.) which are extracted with alkalis. Lignin is found only in mature cells and is made up of an insoluble aromatic polymer resulting from the polymerisation of phenolic alcohols. Suberin consists of Gums and Resins alongwith some aromatic residues and suberic acid.

SUMMARY

Surface architecture of plasma membrane which envelopes the cell; different models of it are discussed. Composition of plasma membrane, functions of the plasma membrane including specialised structures of the cell membrane are described in detail.

EXERCISE:

1. Explain in detail the physical properties and chemical organisation of plasma membrane.
2. What is the unit membrane hypothesis?
3. Explain in detail the chemical composition of plasma membrane.
4. Explain in detail about active transport and passive transport mechanisms.
5. What are the types of transporters of cell membranes? Write in detail.
6. Write short notes on
   (a) Unit membrane model
   (b) Integral proteins
   (c) Ion channels
   (d) Ca^{++} pump
   (e) Microvilli
   (f) Groups translocation
   (g) Vesicles.

3

Cellular Organelles

Like prokaryotic cells, eukaryotic cells are also bound by a plasma membrane and contain ribosomes. However, eukaryotic cells are quite complex and contain a nucleus, a variety of cytoplasmic organelles and a cytoskeleton.

The size of eukaryotic cell varies from 1u to 175000μ (175 mm), The ostrich egg is the largest cell. Nerve cells are very long and some are as long as 3-3.5 feet The number of cells per organism also varies from a single cell in protozoans to millions of cells in higher plants and animals. Unlike prokaryotes, eukaryotic cells are organised into tissues and organs, all of which perform specific functions. Plant cells have a thick complex cell wall whereas the animal cell does not possess any such structure. The plant and animal cells also differ in some other properties. For example plant cells lack centrioles and lysosomes. (Fig. 3.1 and 3.2).

A typical eukaryote contains all the following organelles in its cell.

1. **Cell wall:** It is seen only in plant cells. This is made up of two layers i) Primary Wall ii) Secondary adjacent cells in a tissue have their primary hosts cemented by another layer called middle lamella which is due to pectin, magnesium, calcium which gives rigidity to the wall. The primary wall is thin and elastic and it is made up of pectin and cellulose. The secondary wall has three layers: Outer, Middle and Inner in a concentric manner. In case of sclerenchyma cells the thick secondary wall is made up of number of cellulose bundles called microfibrils.

Such structural details are possible only due to the electron microscope. The electron microscope has enabled a detailed study of tweenfibrils (Micelle) forms and fibrils. Each micelle measures about 250 Å. Each micelle is made up of hundred chains of cellular molecules and 3000 glucose molecules form a cellular chain. To connect the next continuing chain with the protoplasm there are minute perforations called plasmodesmata. Cell wall affords protection, with permeability characteristics, to allow the molecules to pass through metabolic activities of the cell.

2. **Plasma membrane:** It is the outermost covering of animal cells and is present beneath the cell wall in plant cells. It performs functions like transport, synthesis etc., besides giving the cell protection.

3. **Cytosol:** Plasma membrane is followed by the colloidal organic fluid called the cytosol; it responds to a variety of small molecules concerned with metabolism.

4. **Cytoplasm:** Though not an organelle, it is an integral part of any cell and it is in this watery fluid that all the organelles are present and a variety of metabolic reactions takes place.

5. **Golgi bodies:** They are cup shaped, located near the nucleus consisting of smooth, cisternae bounded vesicles which transport proteins.

6. **Endoplasmic reticulum:** It remains continuous with plasma membrane and the nuclear envelope. They are engaged in glycogenolysis and polypeptide synthesis.

7. **Ribosomes:** They act as the work benches of protein synthesis.

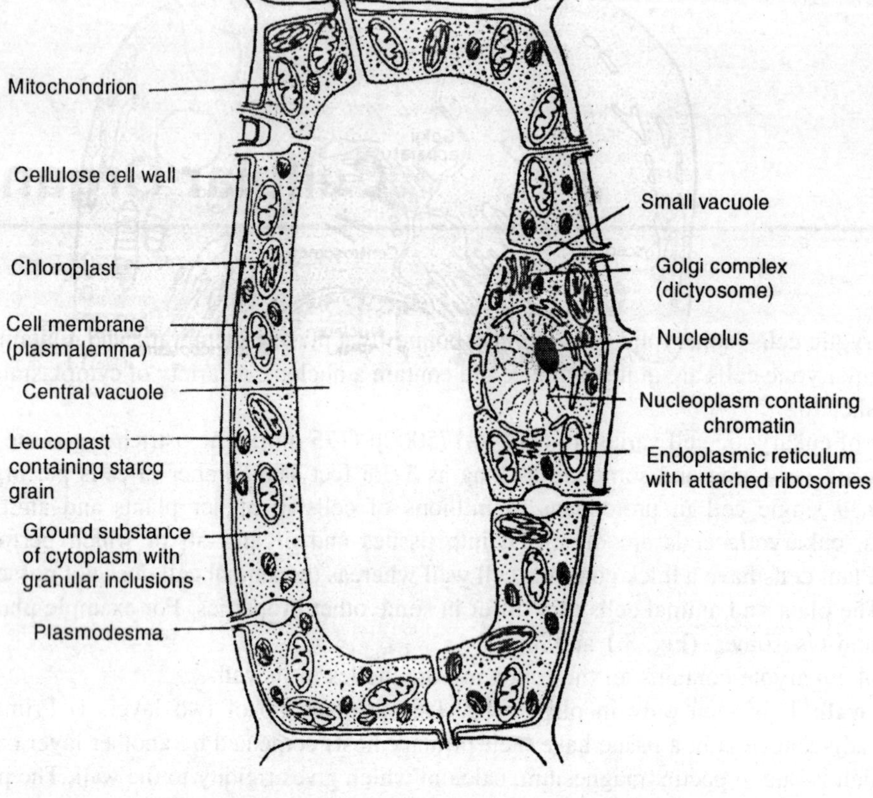

Fig. 3.1. Structure of a typical plant cell organisation

8. **Cytoskeletal structure:** They include fibres and when arranged in 3 dimensional networks are known as microtubular lattice. They maintain cell shape and mobility.

9. **Mitochondria:** They are found in all eukaryotic cells and are the sites of oxidative metabolism. They are responsible for synthesising most of the ATP derived from the breakdown of organic molecules.

10. **Chloroplasts:** They are present only in plant cells and are the sites of photosynthesis.

11. **Lysosomes and peroxisomes:** Also provide specialised metabolic compartments for digestion of macromolecules and various oxidative reactions.

12. **Nucleus:** With a diameter of 5μ which is the distinguishing organelle between the prokaryotic and eukaryotic cell, it is also typical to eukaryotic cells. The nucleus contains DNA arranged in a linear fashion rather than the circular fashion, as in prokaryotes. The nucleus is also the site of DNA replication and RNA synthesis.

CYTOSOL

Cell cytoplasmic organelles can be separated by ultracentrifugation and what is left behind is soluble fraction known as cytosol.

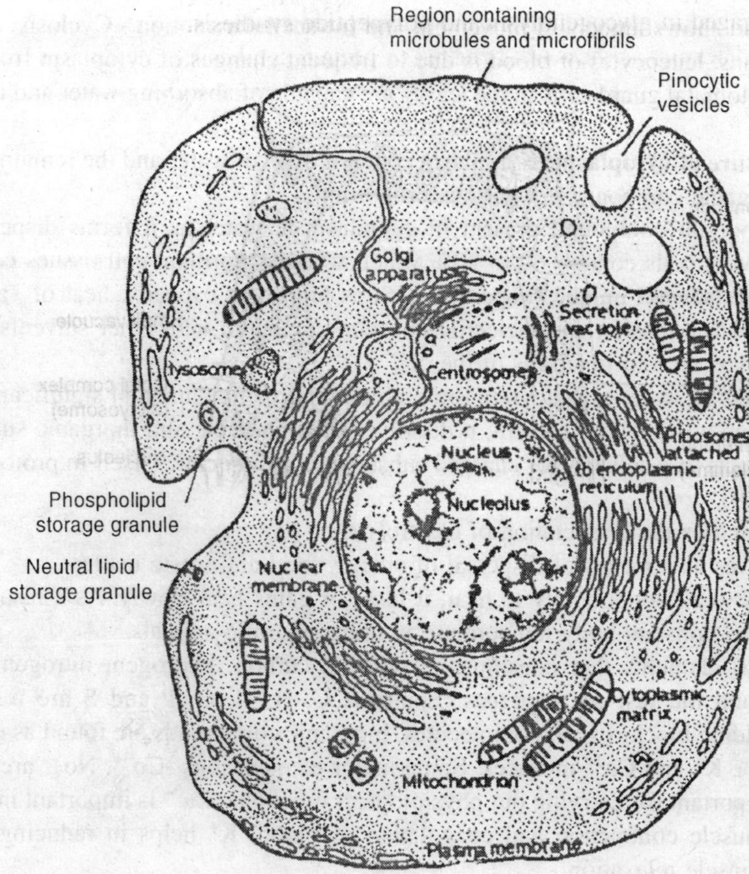

Fig. 3.2. A generalised animal cell organisation

Physical nature of cytoplasm: It appears jelly like, transparent, and colourless fluid. Granier coined a name ergatoplasm for this material. Its basophilic nature is due to the presence of ribonuclease.

Cytoplasm is a heterogeneous mixture of opaque granules and free organic compounds imparting it a colloidal nature. Its ground material has a unique property of changing its consistency, sometimes behaving as a viscous material. The peripheral zone of cytoplasm is thick and jelly like, called plasmagel. Around the nucleus, it is thin-liquefied, called plasmasol or endoplasm. A large number of diverse chemical nature substances are also dissolved in cytoplasm, which has a high percentage of water. Proteins are most abundant, of which 20-25% are in the form of soluble proteins including enzymes. Carbohydrates, inorganic salts, lipids and lipoidal substances are also found in cytoplasm.

Cytoplasm exhibits the phenomenon of phase reversal in which colloidal solution undergoes sol-gel transformation. The sol state is the liquid phase containing more water and less solids whereas the gel state is semi-solid and jelly like, with less water and more solids. Sol gel transformation depends on temperature and occurs in response to metabolic activities of the cell. Cytoplasm exhibits viscosity changes intracellular motion and amoeboid movements.

Cytoplasm suspended particles follow Brownian motion, in which motion of the particles is incoherent. Particles also possess electric charge either negative or positive. As identical charged particles repel, movement occurs.

Cytoplasm also shows amoeboid movement and intracellular motion.- Cyclosis. Amoeboid movement of protozoans, leucocytes of blood is due to frequent changes of cytoplasm from sol to gel state and vice versa. Stomatal guard cells in the leaves are capable of absorbing water and removing it as per the need.

Chemical nature of cytoplasm: Cytoplasm contains 90% of water and the remaining 10% includes organic and inorganic compounds in various proportions.

The most important compound present in cytoplasm is water as it forms dispersion medium of living matter. Living cells contain 60 to 90% water, whereas dormant cell tissues contain 10 to 20% only. As water has several unique properties such as high heat capacity, heat of vaporisation, latent heat of fusion, surface tension and dielectric constant compared with other solvents, these properties influence the behaviour and functions of the cell.

Other properties of water make cytoplasm a compound of fundamental significance like

(i) Water being an excellent solvent, it dissolves many organic and inorganic substances.

(ii) As water acts as a solvent, for charged substances, it dissociates itself in protons and negatively charged OH^- ions.

(iii) Water increases the dissociation of electrolytes.

(iv) Water serves as a means of dispersal of energy, food and waste matter.

(v) Due to its weak conductivity to heat, it loses or gains heat slowly. As a result, water helps in maintaining constant body temperature in homeothermic animals.

Among the 20 elements that constitute cytoplasm, carbon, hydrogen, nitrogen and oxygen are most abundant and the rest are in traces. Elements C, H, N, O, P and S are responsible for the formation of building blocks of all biomolecules. Inorganic compounds are found as ions, such as Na^+, Mg^{2+}, Po^4 2^-. cl^-, K^+ and Ca^{2+} in traces. Some major element ions -Co^{3+}, No_3^- are also present. All these ions are important for carrying out physiological functions. Ca^{2+} is important in causing gelation of cytoplasm, muscle contraction and coagulation of blood. K^+ helps in reducing the viscosity of cytoplasm and muscle relaxation.

GOLGI BODIES

Golgi bodies were identified by Camillo Golgi in 1989. In plants and lower invertebrates it is also called as dictyosome. It is also considered as Golgi complex. The Golgi complex contains parallelly arranged,flattened sac like structures, which lack ribosomes.

Golgi complex or Golgi bodies occur in all cells except the prokaryotic cells and eukaryotic cells of certain fungi, sperm cells of bryophytes and pteridophytes, cells of mature sieve tubes of plants and mature sperm, and red blood cells of animals. Their number in plant cells may vary from several hundred to a single organelle in some algae. In animal cells usually a single golgi complex is present. In oocytes there may be many such bodies. The Golgi body may be scattered throughout the cell or may be generally localised at the place near the nucleus, in the shape of a cup. It gives out several vesicles which later on fuse with the plasma membrane, or form distinct organelles themselves.

Golgi bodies of plant cells are about 1-3μin length and 0.5μ height. It is composed of flattened membrane bound sacs (cisternae) and associated vesicles. Each Golgi complex usually contains about 3-7 flat tubular cisternae. Insects have up to 30 cisternae. The Golgi body has a distinct polarity in both structure and function. The face of Golgi body faces the Endoplasmicreticulum and is usually convex and is called the cis-face or the forming phase. The proteins from the Endoplasmic reticulum enter the Golgi body from this side. The opposite side which is concave is called the transface or the maturing face from which the proteins exit. The Golgi complex consists of three functioning distinct compartments. The cis-Golgi network, the golgi stack and the trans-golgi network. Transport vesicles carry the proteins between these networks. In addition,membrane tubules may also connect the Golgi compartments.

A Golgi body is surrounded by a differentiated region of cytoplasm, where ribosomes, glycogen and other organelles are scarce or absent. This zone is called the zone of exclusion. In conjunction with the endoplasmic reticulum, lysosomes etc. the Golgi body forms an extensive inter communicating membrane system called the endomembrane system. This forms an exclusive area inside the cell, where various processing mechanisms and transport continuously occur, separated from the cytoplasm. The Golgi body is majorly involved in transport, processing of proteins, other than synthesizing polysaccharides and cell wall components in plants (Fig. 3.3).

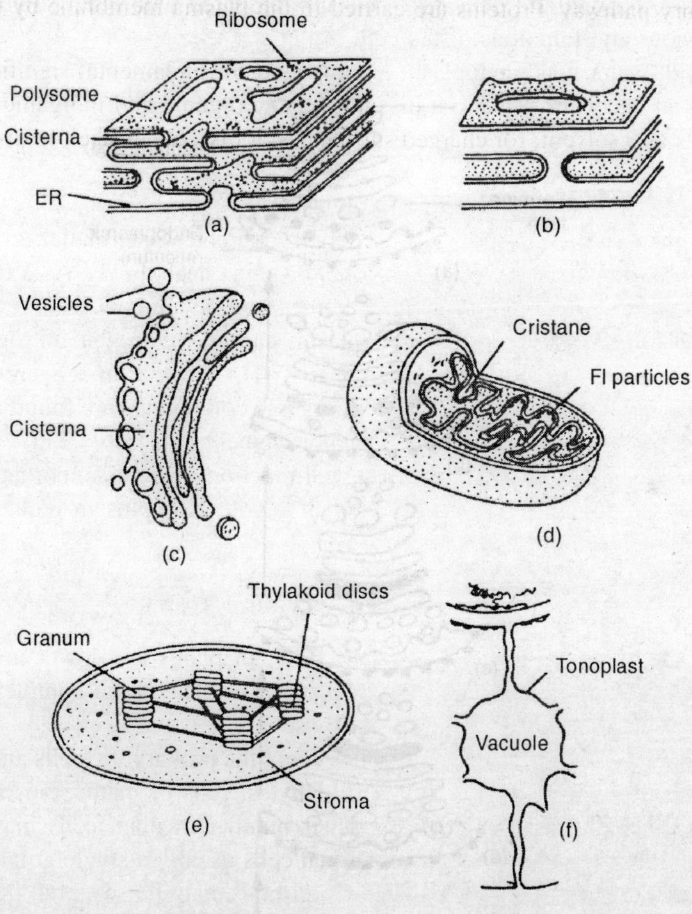

Fig. 3.3. Various cell organelles found in cells (only some important ones given) (a) Granular endoplasmic reticulum (b) Smooth endoplasmic reticulum (c) Golgi complex

Functions: Proteins enter the Golgi from the cis-golgi network which mainly serves to receive the transport vesicles from the Endoplasmic reticulum. Proteins marked to be present within the Endoplasmic reticulum are recognised and returned to the Endoplasmic reticulum by recycling pathway with the help of receptors and signals. Other proteins are transported to the Golgi stack, which is the site of most metabolic activities of the Golgi apparatus. The proteins are processed here and then transported to the trans-golgi network where the final stages of processing are completed. The trans-golgi network acts as a sorting centre, directing the proteins and other molecules to the lysosomes, plasma membrane or the cell exterior. Another important function is to make the N-linked oligosaccharide that has to be added in endoplasmic reticulum. The proteins targeted for lysosomes are added with mannose-6-phosphate residue as the last step before which several sugars are removed. In the Golgi body carbohydrates are added to the serine or threonine residues (Fig. 3.3a).

Golgi bodies also participate in lipid metabolism, i.e. synthesis of glyco-lipids and sphingomyelin.

Plant cell Golgi bodies serve as a site where complex polysaccharides are synthesised. They are used in cell wall formation. Hemicellulose of pectin are synthesised in the Golgi bodies and transferred to the cell surface by transport vesicles.

Proteins, lipids and polysaccharides are transported from Golgi body to their final destinations, through the secretory pathway. Proteins are carried to the plasma membrane by bulk flow.

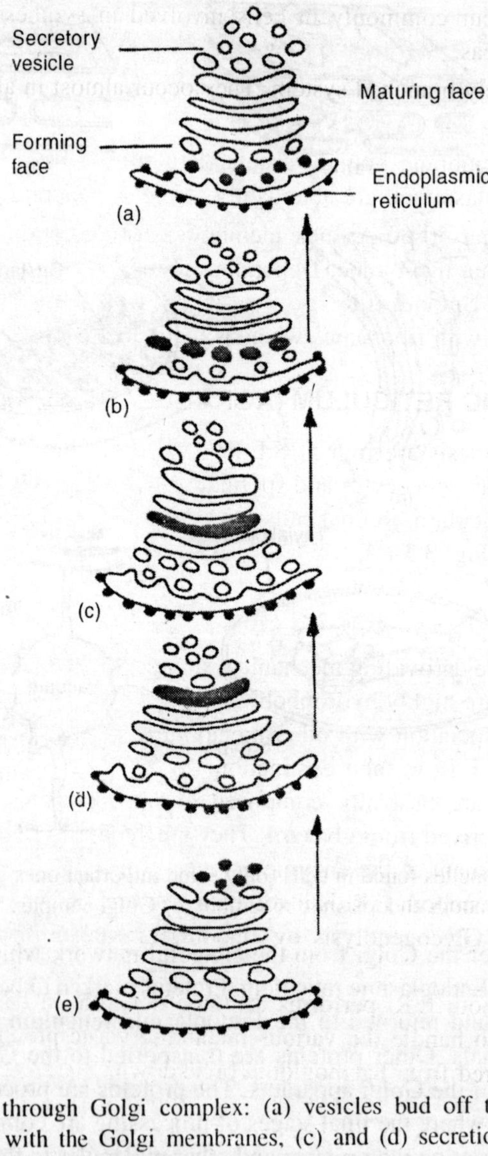

Fig. 3.3a. Membrane flow through Golgi complex: (a) vesicles bud off the endoplasmic membrane, (b) vesicles fuse with the Golgi membranes, (c) and (d) secretions move through the Golgi cisternae, and finally (e) released

In yeasts and plant cell, proteins are transported from the Golgi body to the vacuole as they lack lysosomes. Golgi body contains various enzymes viz., phospholipase ADPase, ATPase, CTPase, Thyamine pyrophosphatase, galactose transferase, Glucose-6-Phosphotase, etc.

Holocrine, apocrine and merocrine like cellular secretions are caused due to Golgi bodies.

ENDOPLASMIC RETICULUM (ER)

The name Endoplasmic Reticulum was coined by Porter (1953) after observing in liver cells. It is a network of tubules and sacs which extend from nuclear membrane throughout the cytoplasm. Cells involved in protein synthesis and storage contain large amount of ER, whereas erythrocytes, (RBC) eggs and embryonic cells lack ER.

ER consists of cisternae, vesicles and tubules.

(a) **Cisternae** are long flattened sac like structures with 40-50 mμ in diameter. They are arranged in bundles paralelly. They occur commonly in cells involved in synthesis of various substances. Ex. brain, notochord and pancreas.

(b) **Tubules** are reticulately branched system. They occur almost in all cells with a diameter of 50-500 mμ.

(c) **Vesicles** are oval membrane bound vascular structures with a diameter of 25-500 mμ they remain isolated in the cytoplasm and are abundantly found in pancreas.

Structurally, it is trilaminar 50-60 Å thick membrane having continuation with the nuclear membrane Golgi body and plasma membrane. Depending upon their surface morphology, E.R. is of two types, viz., Rough E.R. and Smooth E.R.

The rough E.R is coated with ribosomes whereas the smooth E.R. is devoid of ribosomes.

SMOOTH ENDOPLASMIC RETICULUM (AGRANULAR) (SER)

Cells involved in lipid synthesis are rich in S.E.R. Also it is found in glycogen storing liver cells, adipose cells, interstitial cells, leucocytes and spermatocytes. Muscle cells also possess SER; here it is called the sarcoplasmic recticulum. Retinal cells possess them as tightly packed vesicles and hence are called the myeloid bodies (Fig. 3.3).

FUNCTIONS

(1) **Synthesis of fats:** Besides providing mechanical support to the cell the SER is actively involved in lipid synthesis. As the fats are highly hydrophobic, they are synthesised in the cytosol. Although some lipids are synthesised in association with other membranes, most of them are synthesised at SER and are transported through the E.R. to their destination.

Eukaryotic membranes are basically composed of phospholipids, glycolipids and cholesterol. Phospholipids are mainly derived from glycerol. They are synthesised on the cytosolic side of the E.R. membrane with the involvement of the enzyme glycerol-3-phosphate and coenzyme A.

(2) **Glycogenolysis:** The smooth E.R. is involved in the breakdown of glycogen, through the action of glucose-6-phosphotase. Glycogenolysis by smooth E.R. takes place only in the cells of fasted animals.

(3) **Detoxification:** Smooth E.R. performs an important role in detoxification in liver cells. The detoxification reactions also handle the various metabolic waste products like fatty acids, bile salts, steroids and haeme recovered from haemoglobin breakdown.

4) **Cholesterol metabolism:** S.E.R. also plays a role in cholesterol and lipid metabolism in the liver and secretion of CL^- ions in the HCL secreting cells of the stomach, secretion of triglycerides in the intestinal absorptive cells, synthesis and secretions of steroid hormones by mammalian gonads. The pigments of the retinal cells are also synthesised by S.E.R. from vitamin A.

GRANULAR OR ROUGH ENDOPLASMIC RETICULUM (RER)

Endoplasmic reticulum coated with ribosomes is called the rough E.R. In the cells which are involved in active protein synthesis, such as plasma cells, pancreatic cells, endocrine glands cells and liver cells, E.R. is coated with ribosomes extensively, as these ribosomes take up basic dyes and hence are termed as basiphilic bodies, Nissls bodies etc. Ribosomes are work benches of protein synthesis (Fig. 3.3).

FUNCTIONS

(1) **Sorting of protein and secretion:** Lysosomes or Plasma membrane are synthesised on the membrane bound ribosomes which enter the E.R. for being sorted and secreted.

The m-RNA'S to be translated on membrane bound ribosomes contain a set of unique codons just 3′ of the initiation site and translation of these codons yields a unique sequence at the amino terminal of the growing polypeptide called the signal sequence, which triggers attachment of the ribosome to the membrane.

Protein processing is related to the E.R. and the 20 amino acid sequence is not necessary for the functioning of the protein and is involved in tagging the polypeptide to the E.R.

Transmembrane proteins are translocated partially into the lumen of E.R. Further translocation is stopped by a stop-transfer sequence on the polypeptide. The lumen of E.R. contains various enzymes and proteins which are involved in proper folding and functioning of proteins.

(2) **Glycosylation of proteins:** Proteins which enter E.R. are glycosylated on specific aspargine residues on the protein, while their translation is still in process. Glycosylation involves the addition of an oligosaccharide to the aspargine residues. A membrane bound enzyme called oligosaccharyl transferase transfers the oligosaccharide to the Aspargine. Further synthesis of glycerides, fatty acids and fatty acids biosynthesis takes place in E.R. membrane due to the involvement of various enzymes.

COMMON FUNCTIONS OF BOTH S.E.R. AND R.E.R.

(1) An ultra structural framework to the cell is provided.
(2) Transport by permeases and carriers like in plasma membrane.
(3) Increases the area of enzymatic reactions.
(4) Different proteins are secreted from the E.R. to Golgi apparatus and then to secretory vesicles. The proteins destined to E.R. are initially sent to Golgi appratus and then again brought back to the E.R. with the help of specific receptors and signals added to it.
(5) They organise intracellular signals.
(6) E.R. forms new nuclear envelope after cell division.
(7) Sarcoplasmic reticulum in muscle cells is involved in Ca^{++} ions releasing during muscle contraction.

RIBOSOMES

Ribosomes are small globular organelles of 250 Å diameter. Robinson and Brown in 1953 have reported, firstly, ribosomes in bean roots. In 1956 Palade, identified ribosomes from Amphibian oocytes and discovered the presence of RNA in ribosome. Chloroplasts mitochondria and nucleus also possess ribosomes. Ribosomes or ribonucleo protein particles are situated either on the wall of endoplasmic reticulum or are present in cytoplasm independently.

It is estimated that mammalian cells contain 10 million ribosomes per cell and in E.coli 10,000 - 20,000 per set. Ribosomes are many in actively metabolising cells. They have 2 subunits and are of 2 types.

(a) *70s Ribosomes:* are smaller with a sedimentation coefficient of 70s. They are found in Bacteria, Mitochondria and Chloroplast.

(b) *80s Ribosomes:* are larger in size with a sedimentation coefficient of 80s found in cells of higher plants and animals.

Each of these is consisting of two subunits, one larger and other smaller subunit. 70s one is coupled with a 50s subunit and a 30s subunit. 80s subunit is coupled with 60s and 40s.

S = Sved berg units (Sedimentation coefficient)

However, certain exceptional ribosomes are seen in mammalian mito-chondria, consisting of 35s and 25s subunits. Yeast possesses 73s ribosomes and fungi possess 77s ribosomes in their mitochondria.

Subunits of ribosomes are associated with each other only during protein synthesis. $Mg2^+$ concentration of 0.001M in cytoplasm is favourable for subunits association.

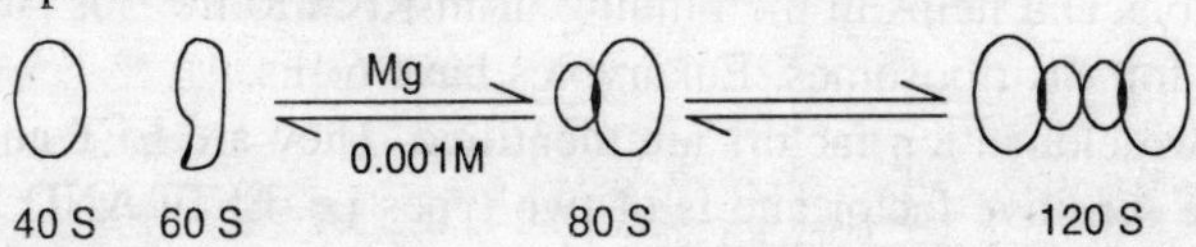

Figure showing Mg^{++} concentration dependent association of ribosomes or subunits.

Chemical composition: Proteins and RNA are the composition of ribo-somes without any lipid content.

Polysomes: Ribosomes are often found in groups; such conditions are considered as polysomes or polyribosomes.

Structure: The 40s and 30s subunits are rather curved and consist of a head, a base, cleft and a platform. The larger subunit is either triangular or rounded in shape. It has two concave sides and one flattened surface.

The cleft of smaller subunit lodges m-RNA and a small gap between the two subunits accommodates t-RNA to bind to it. The concavity of larger subunit accommodates polypeptide.

Ribosomal RNA and proteins: RNA and Proteins are major constituents of Ribosomes. Ribosomal RNA (rRNA) are present in more than 80% of the RNA present in cells.

Prokaryotic ribosomes possess 3RNA molecules whereas in Eukaryotes, there are 4 rRNAs.

The rRNA contains several nucleotides as follows:

23sr-RNA = 3300 Nucleotides.
16sr-RNA = 1650 "
5sr-RNA = 120 "

Protein ratio differs from Prokaryotic ribosome to Eukaryotic ribosome.

E.Coli ribosome possess 40-60% rRNA. and 37% Protein

Eukaryotic ribosome possess 40-44% rRNA and 60% Protein.

In 1960, the role of ribosomes in protein synthesis was enlightened. It is considered that the rRNA provides a structural background for the proteins to catalyse different reactions; further the comparisons with primitive self catalytic RNA (Ex.RNA-ase-P and Ribozymes) led to prove the catalytic role of RNA.

FUNCTIONS

(A) Ribosomes are involved in protein synthesis; specific function being codon and anticodon recognition between t-RNA and m-RNA. This is brought about by r-RNA. r-RNA also participates in protein synthesis due to its base-pairing quality.

(B) To form the polypeptide chain, peptide bond formation between the amino acids.

(C) Translocation is performed with the help of protein factors and certain enzymes viz.,
  (i) Initiation factors,
  (ii) Elongation factors,
  (iii) Release factors.

INITIATION FACTORS

They are three – IF1, IF2 and IF3 according to Ochoa, Stanley and Iwanowski (1968).

IF1: It weighs 92K daltons and helps in the binding of formyl methionine tRNA to 30s subunits.

IF2: It weighs 80K daltons. It has an SH group associated with it, which regulates the binding of GTP.

IF3: It weighs 30K daltons and helps in the binding of m-RNA to the 30s subunit. It also acts as a dissociation factor for 70s ribosomes. Eukaryotes have 6 IFs.

Elongation factors: Two elongation factors are identified. They are EFT and EFG.

(1) EFT is temperature sensitive factor and is of two types i.e. EFTu AND EFTs.

EFTu: Elongation factor temperature unstable; it helps in the transfer of amino-acetyl tRNA to the A-site (GTP is converted to GDP)

EFTs: Elongation factor temperature stable: Its function is to activate the inactivated EFTu-GDP to EFTu - GTP (GDP is converted to GTP)

(2) EFG is also called as G-factor and also as translocase as EFG along with GTP helps in the movement of ribosome on the mRNA (Trans-location) during protein synthesis.

Release factors: 2 release factors RF1 and RF2 are available in prokaryotes. They identify the termination codons UAA, UAG and UGA and help in the termination of polypeptide chain.

Enzymes: An enzyme known as peptidyl transferase is the 50s ribosomal subunit, which helps in the formation of peptide bonds. This enzymes nature is similar to that of ribozymes, i.e. catalytic RNA along with some proteins.

Cytoskeletal structure: The term cytoskeleton is applied to the microtubules, microfilaments and also intermediate filaments spread through the cytosol. Tubulin (in Microtubules) Actin, Myosin, Tropomyosin (in Microfilaments) are the main proteins present in the cytoskeleton.

Microtubules: De Robertis and Franchi in 1953 have identified microtubules in nerve cells, and were called as neurotubules or nerve tubules. Sabatini, Bansch, Barnette in 1963 have explained actual structure of microtubules. Ledbetter and Porter have found microtubules in plant cells.

Microtubules are found in Cilia, and flagella, basal granules, Centrioles, Kinetochore, polarcaps in plants, spindle fibres; axoneme in protozoa, cortex of meristematic cells in plants and elongated cells like lens cells and sperms.

Structure: Mitrotubules are long, unbranched and hollow, completely made up of proteins. In length they range several micrometers and in dia 210-250 Å. Microtubule is composed of protein sub units that are globular and are called "tubulin protein".

Two different monomers, Tubulin-a and Tubulin-b are present. The a and b subunits are associated as dimer pH to form the microtubule depends upon pH, temperature, cation concentration and GTP. It was found that microtubules assemble at 37°C. Mg++, Ca++ are also necessary for the association of these units to form a microtubule. Besides Tubulin protein, microtubules also contain MAP - Microtubule Associated protein, which is involved in microtubular assembly.

FUNCTIONS

Microtubules help in movement of Cilia, and Flagella, Cyclosis or Cytoplasmic streaming, maintaining the cell structure movement of chromosomes during cell division, the division of the cell during cytokinesis. Microtubules involve in shaping the cell during cell differentiation. Further they help protozoans to procure their food, transport of particulate matter, including determination of intrinsic polarity in certain cells.

Microfilaments: Microfilaments consist of Actin Filaments which are smaller in structure than microtubules and very active. They were first observed in muscle cells in 1947.

Actin filaments consist of monomers. They are proteinaceous in nature and each actin monomer is known as G-Actin (Globular Actin) which polymerise to form F-actin (Fillamentous actin). Each G-actin is made up of 375 amino acids. Association and Dissociation is dependent upon Cation concentration. Polymerisation of monomers is also helped by Actin binding proteins. The actin monomers attached to ATP residues mainly attack at +ve end. While the actin monomers with ADP residues are no residue remain at the -ve end. There is a rotation of 160° in between one molecule and other when an actin filament is formed which results in the formation of double stranded helix like appearance.

Intermediate filaments: They are unbranched structures having a diameter of about 100Å, which are intermediate between microtubules and micro-filaments. These are not directly involved in the cell mobility, but give mechanical strength to the cell. More than 50 types of intermediate filaments have been identified. They have been divided into 6 types.

(1) Acidic Keratin (epithelial cells)

(2) Neutral Keratin and Basic Keratin (epithelial cells)

(3) Vimentin in (WBC and Fibroplast)
Desmin (Muscle cells)
Peripherin (Peripheral nervous system)

(4) Neurofilament proteins (In nerves)

(5) Nuclear lamins (Nuclear lamina)

(6) Nestin (Central Nervous System)

MITOCHONDRIA:

Mitochondria were first described by Altmann in 1890 as Bioblasts. The term Mitochondrion was coined in 1897 by Benda. They are commonly referred to as power houses of the cell and specialize in the synthesis of adenosine triphosphate (ATP), the useful chemical form of energy.

They are present in all aerobic eukaryotic cells except in RBC. They are uniformly distributed in the cytoplasm. In some cases they are located near the nucleus. In general, mitochondria are located near such structures where energy requirements are heavier.

Structure: Mitochondria are surrounded by a double membrane system consisting of an inner and outer mitochondrial membrane separated by an inter membrane space. The inner membrane forms numerous folds called cristae which extend into the interior of the matrix.

Outer mitochondrial membrane is freely permeable to small molecules. It contains proteins called Porins, which form channels that allow the free diffusion of molecules smaller than about 6000 daltons. Composition of the intermembrane space is similar to the cytosol with respect to ions and small molecules (Fig. 3.4).

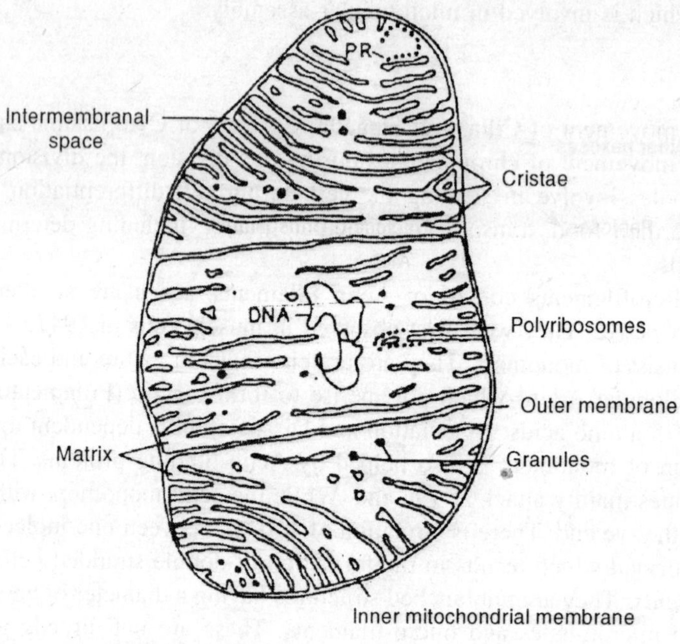

Fig. 3.4. Diagrammatic presentation of a mitochondrium showing typical internal structure

Mitochondria contain their genetic system, which is separate and distinct from the nuclear genome of the cell. Mitochondrial genomes are usually circular.

Functions: Oxidative phosphorylation (Fig. 3.4a):

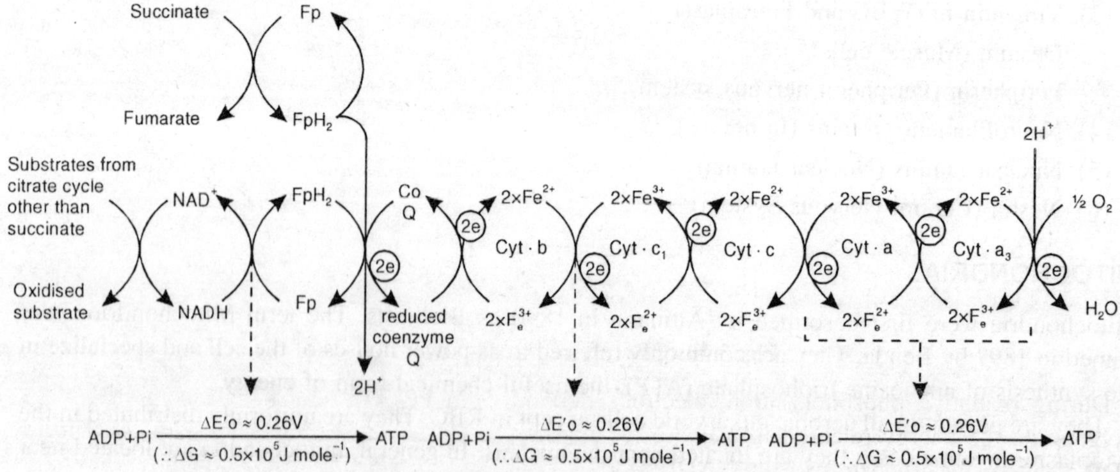

Fig. 3.4a. Electron transport chain and the locations where ATP is synthesised

Majority of the usable energy obtained from the breakdown of carbohydrates or fats is derived by oxidative phosphorylation which occurs within the mitochondria. The breakdown of glucose by glycolysis and citric acid cycle yields a total of four molecules of ATP and ten molecules of NADH and two molecules of FADH2. They are then transferred to molecular oxygen, coupled to the formation of an additional 32 to 34 ATP molecules by oxidative phosphorylation. Electron transport and oxidative phos-phorylation are critical activities of the protein complexes present in the inner mitochondrial membrane (Fig. 3.4b).

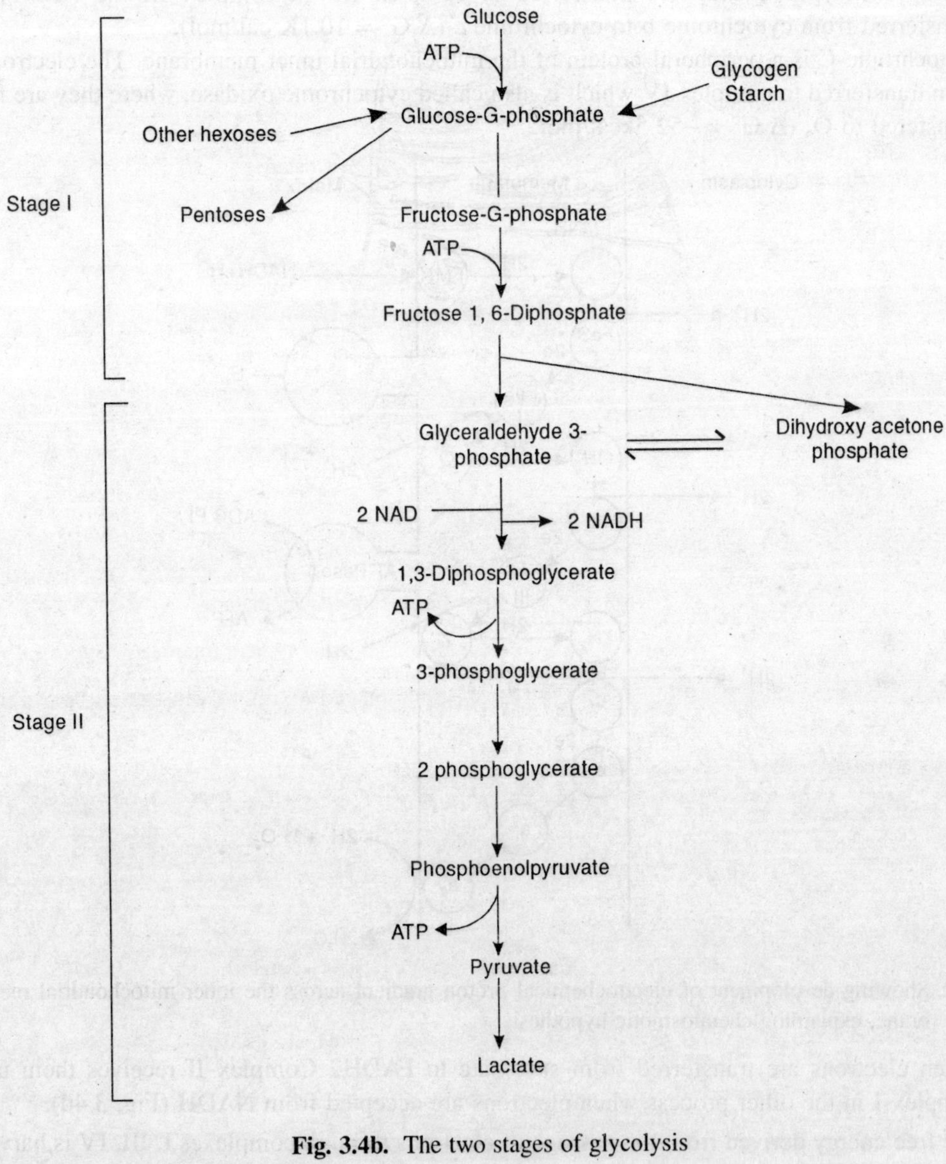

Fig. 3.4b. The two stages of glycolysis

During oxidative phosphorylation, electrons derived from FADH2 and NADH combined with oxygen and the energy released from these reaction is utilised to derive the synthesis of ATP from ADP. The transfer of electrons from NADH to O_2 is a very energy yielding reaction with $\Delta G° = 52.5$

Kcal/mol for each pair of electrons transferred to be harvested in a usable form; the energy must be graudally produced by the passage of electrons through a series of carriers which constitute the electron transport chain (Fig. 3.5).

The steps of electron transport are as follows:

(1) Electrons from NADH enter the electron transport chain in complex I. These are first transferred to flavin mononucleoxide (FMN) and these form an Iron Sulfer carries to co-enzyme-Q (Ubiquinone) Δ G° = 52. 5Kal/mol. (Fig. 3.4c).

(2) From co-enzyme Q they are transferred to complex III. In complex III the electrons are transferred from cytochrome b to cytochrome c (Δ G° = 10.1K cal/mol).

(3) Cytochrome-C is a peripheral protein of the mitochondrial inner membrane. The electrons are then transferred to complex IV which is also called cytochrome oxidase, where they are finally transfered to O_2 (Δ G° = –52.5kcal/mol.

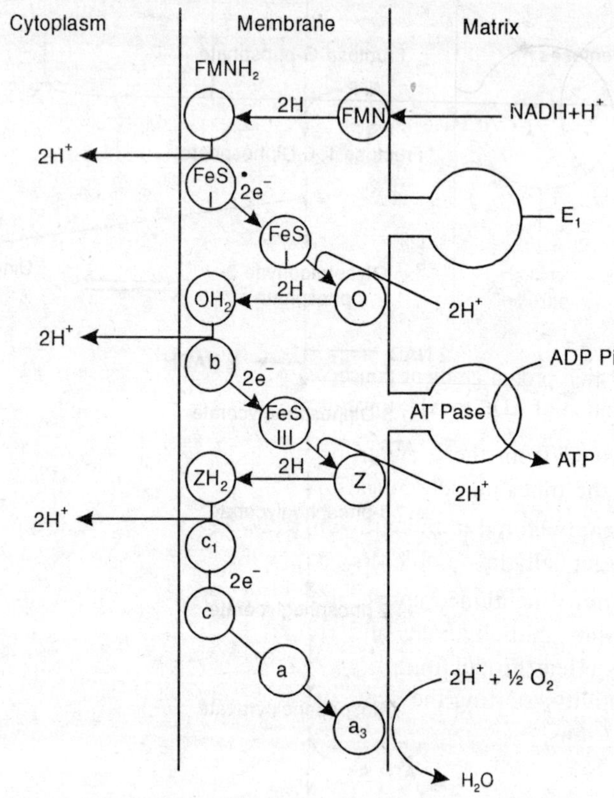

Fig. 3.4c. Showing development of electrochemical proton gradient across the inner mitochondrial membrane, explaining chemiosmotic hypothesis

(4) When electrons are transferred from succinate to FADH2 Complex II receives them unlike complex I in the other process when electrons are accepted from NADH (Fig. 3.4d).

(5) The free energy derived from the passage of electrons through complexes I, III, IV is harvested by being coupled to ATP synthesis.

(6) Here the ATP is generated by the directed phosphorylation of ADP to ATP as in cases of glycolysis, etc. (Fig. 3.4e).

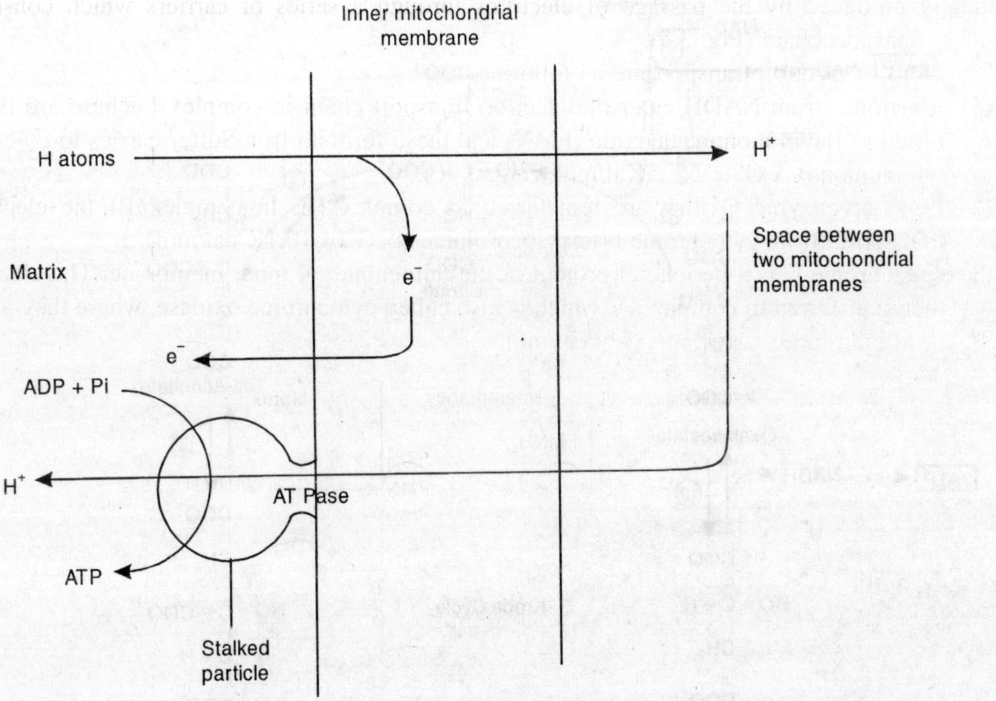

Fig. 3.4d. The energy-rich proton gradient conserves free energy of electrons and ultimately accounts for phosphorylation of ADP to ATP towards the matrix

(7) The energy derived from these electrons transport is coupled to the generation of a proton gradient across the inner mitochondrial membrane.

(8) The potential energy stored in this gradient is then harvested by a fifth protein complex which couples the energetically favourable flow of protons back across the membrane to ATP synthesis.

Other than the functions mentioned above, mitochondria also perform the following functions:

(1) Degradation of fats, carbohydrates and proteins.

(2) Thermiogenesis (Heat Production).

(3) Biosynthetic activity (starting the synthesis of Haem etc).

(4) Storage of Ca^{++} ions.

LYSOSOMES:

Lysosomes are tiny, membrane bound vesicular structures of the cell which possess destructive enzymes called hydrolases, which are capable of breaking down macromolecules. Deduve in 1955 reported about lysosomes.

Lysosomes occur in all mammalian cells except in mature RBC. Many plant cells also are devoid of it. Pancreatic cells, liver cells, leucocytes, spleen cells have large number of lysosomes. Macrophages, too, have large number of lysosomes.

Lysosomes are spherical in shape. Size varies from 0.2 to 0.8μm; in kidney cells their size is 5μ; they are larger in phagocytes and leucocytes. They are round vacuolar structures filled with dense material including acid hydrolases and phosphatases. They are bounded by a unit membrane (Fig. 3.6).

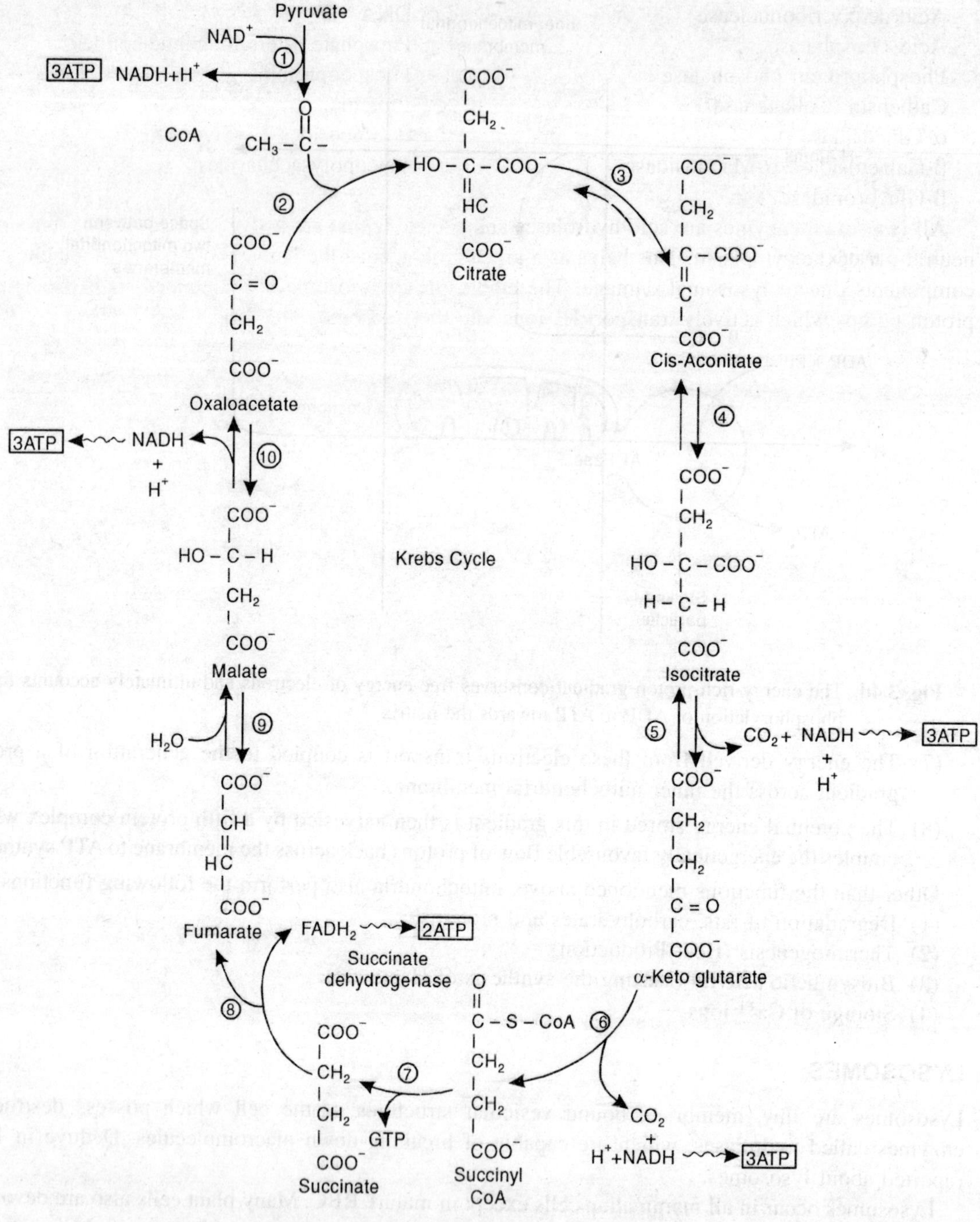

Fig. 3.4e. The tricarboxylic acid cycle (Krebs cycle)

Lysosomes are found to contain 40 different enzymes or hydrolases. Some of the enzymes and their substrates are listed below:

Enzymes	Substrates
Acid ribonuclease	RNA

Acid deoxy ribonuclease	DNA
Acid Phosphatase	Phosphate esters mononucleotides
Phosphoprotein phosphatase	Phosphoproteins
Cathepsin, Collagenase	Proteins
α-Glucosidase	α-Glucosides ← Glycogen
β-Galactosidase α-Mannosidase }	Mucopolysaccharides
β-Glucoronidase }	

All lysosomal enzymes are acid hydrolases and proteases that are active at pH5 (acidic), but not at neutral-pH of the cytoplasm. This helps as a precaution against the damage of cellular organelles and components due to lysosomal damage. The acidic pH of lysosome is maintained with the help of proton pumps which actively transport H^+ ions into the lysosome.

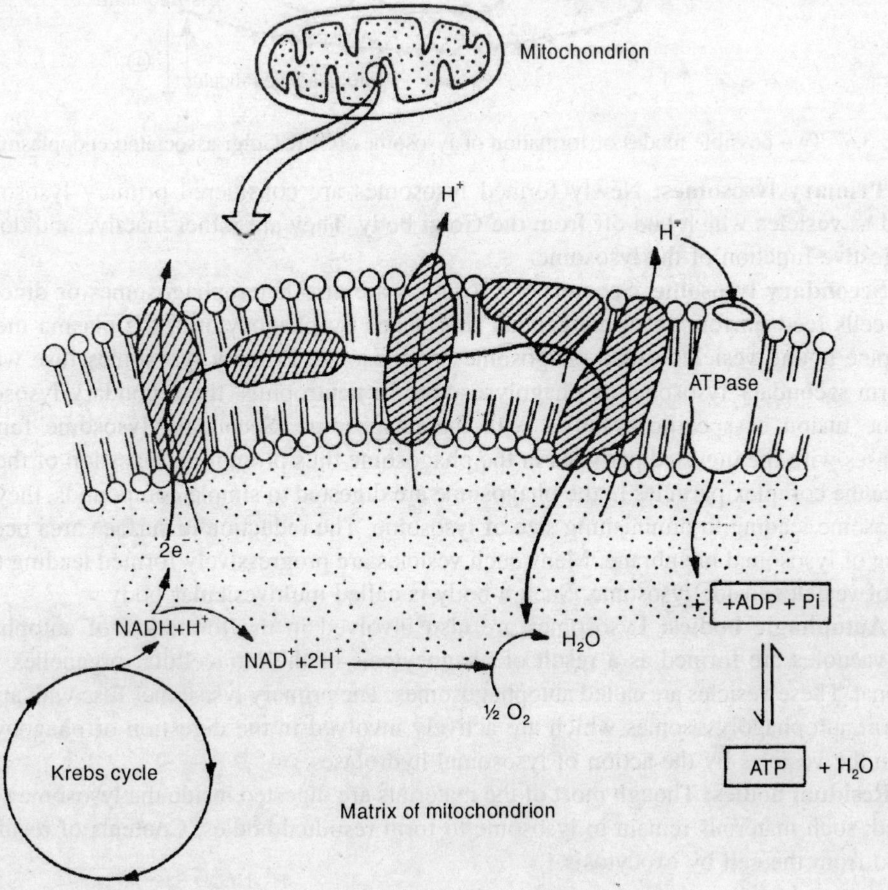

Fig. 3.5. Oxidative phosphorylation in mitochondrial inner membrane

KINDS OF LYSOSOMES

Different cells may have different lysosomes or the same cell may have different forms of lysosomes during different phases of development. They are identified as 4 types.

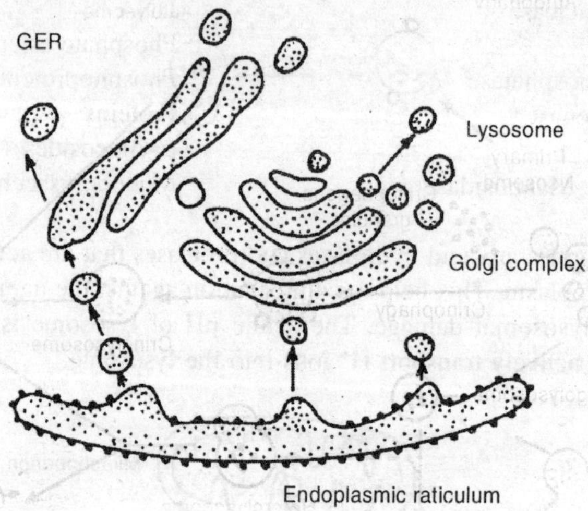

Fig. 3.6. Two possible modes of formation of lysosomes (GER Golgi-associated endoplasmic reticulum)

(1) **Primary lysosomes:** Newly formed lysosomes are considered primary lysosomes. They are formed as vesicles which bud off from the Golgi body. They are rather inactive and do not involve in the digestive function of the lysosome.

(2) **Secondary lysosome:** Secondary lysosomes are also heterophagosomes or digestive vacuoles. When cells feed on foreign or exogenous substances by phagocytosis, the plasma membrane forms membrane bound vesicles called phagosome or pinosome. Primary lysosomes fuse with phagosome and form secondary lysosome or phagolysosome. In neutrophils, the secondary lysosome is formed after the fusion of specific granules with the phagosome. Secondary lysosome functions to mix hydrolases with the ingested particles in the phagosome thus promoting digestion of these substances.

Once the complex particles in the phagosome are digested to simple compounds, they diffuse out of the lysosome leading to diminishing size of lysosome. The reduction in surface area occurs by inward budding of lysosomal membrane. Many such vesicles are progressively formed leading to the accumulation of vesicles inside lysosome. Such a body is called multivesicular body.

(3) **Autophagic bodies:** Lysosomes are also involved in the formation of autophagic vacuoles. These vacuoles are formed as a result of phagocytosis of the intracellular organelles which are non functional. These vesicles are called autophagosomes. The primary lysosomes fuse with autophagosomes and form autophagolysosomes which are actively involved in the digestion of phagocytosed components in the vesicles by the action of lysosomal hydrolases.

(4) **Residual bodies:** Though most of the materials are digested inside the lysosome some cannot be digested; such materials remain in lysosome to form resudual bodies. Contents of residual bodies are released from the cell by exocytosis.

FUNCTIONS

(1) Digestion of large extracellular particles by the formation of phagolysosomes. They are also responsible for killing of bacteria and viruses that are phagocytosed, by fusion with phagosome (Fig. 3.7).

(2) During starvation lysosomes are seen actively involved in the digestion of intracellular organelles for energy.

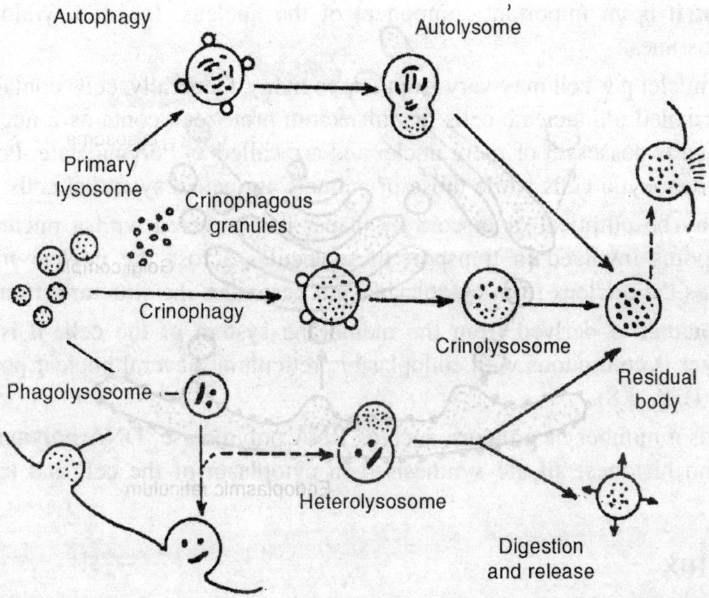

Fig. 3.7. Cellular digestion in lysosomes showing crinophagy and formation of residual bodies

(3) Autolysis also occurs with the help of lysosome. In insects and amphibian larvae, during metamorphosis loss of various body parts occurs by lysosomal action. This process is called autophagy and lysosomes are commonly considered suicide bags of the cell.

(4) Extracellular digestion by the secretion of lysosomal enzymes in some gametes like sperm also done by lysosomes.

(5) Lysosomes are also supposed to initiate mitosis in cells.

LYSOSOMES AND DISEASES

As a result of defects in normal lysosomal function various diseases occur.

Hydrogen peroxide is an important chemical in phagocytic killing and the disease is called chronic granulomatic disease. This character is explained by an X-linked recessive condition. In this disease due to absence of H_2O_2 in lysosomes, killing of bacteria and fungi also do not occur. Subsequently many macrophages just phagocytose bacteria and retain them within themselves and accumulate in an area. This condition is called Granuloma and the disease is termed "Chronic granulomatous diseases".

Gaucher's disease occurs where lysosome lacks an enzyme "gluco-cerebrosidase" which cleaves glucocerebroside to glucose and ceramide. Due to the deficiency the substrate i.e. glucocerebroside, accumulates inside the lysosome causing disease.

Diseases also occur when lysosome fails to fuse with phagosome. At this time bacteria present in phagosome divide and multiply and cause disease. The diseases like Gaucher's disease which result from the deficiency of lysosomal hydrolasers are called lysosomal storage diseases as they result in accumulation of a particular product in lysosome.

NUCLEUS

The nucleus was discovered by Robert Brown in 1831. The nucleus is considered to be the core of the cell. It commands all metabolic events. Miescher (1869) discovered nucleoproteins and nucleic acids and suggested that it is an important component of the nucleus. In 1888, Waldeyer observed and designated chromosomes.

The number of nuclei per cell may vary from one to many. Generally, cells contain a single nucleus, and such cells are called uninucleate cells. Paramaecium protozoan contains 2 nuclei and is called as Binucleate. Some cells possess 3 or more nuclei and are called as Polynucleate. Polynucleate cells of plants are called coenocytic cells while those of animals are called syncytial cells.

The nucleus may be elliptical or discoid in shape; it is covered with a nuclear envelope which contains nuclear pores involved in transport of molecules across the nuclear membrane. Nuclear membrane separates the nucleus from cytoplasma and peroxides the structural framework of nucleus.

Nuclear membrane: is derived from the membrane system of the cell. It is a double layered structure; outer layer is continuous with endoplasmic reticulum. Several nuclear pore complexes exist on this membrane (Fig. 3.8).

Nucleus contains a number of proteins, such as RNA polymerase, DNA polymerase, several types of cytochromes and histones; all are synthesised in cytoplasm of the cell and transported into the nucleus.

NUCLEAR MATRIX

Within the nuclear envelope the nucleus body possesses a dense jelly like mass composed of karyolymphs and chromatin. Karyolymph is rich in proteins with a small percentage of DNA, RNA and Phospholipids. In this matrix DNA replication, transcription and transport of substances take place. Nuclear metrix has a network of interchromatinic matrix.

There is a nucleolus in the nucleus; its number may vary in different cells. Cells lacking nucleolus show little or no protein synthesis. Nucleolus or nucleoli are situated along certain chromosomes. The morphology of nucleolus undergoes change during cell cycle. During prophase it disappears and during Metaphase it reappears.

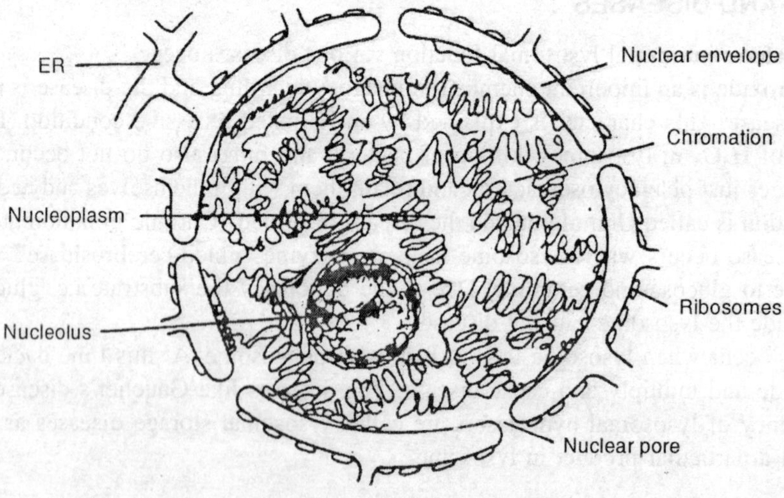

Fig. 3.8. Nucleus of eukoryotic cell

FUNCTIONS OF NUCLEOLUS

It is the most important structure in nucleus, being the site of rRNA transcription and processing of ribosome assembly.

It possesses genes for the 5s, 58s, 18s and 28s tRNA. The 58S, 18S and 28s, rRNA are transcribed as a set and the 5srRNA is transcribed out of the nucleolus. There are around 2000 'copies of the genes that code for 5.8s, 18s and 28srRNA and around 2000 copies of the genes that code for 5s rRNA.The 5.8s,18s and 28s rRNA. genes are distributed on five different chromosomes in humans (13, 14, 15, 21 and 22) and 5s genes are present on chromosome No.1. Repeated ribosomal genes serve to produce a large number of ribosomes required for protein synthesis.

Nucleolus does 3 major functions i.e. (i) Transcription of the genes that code for ribosomal RNA. ii) Processing of periribosomal molecule and iii) Assembly of ribosomal subunits.

The nucleolus organiser region of the chromosome consists of multiple copies of the genes that code for rRNA, some of which are redundant. The nucleolar genes produce rRNA, hence they are called ribosomal genes which are rich in nucleotides containing guanine and cytosine.

The 45s rRNA is methylated, cleaved and reduced to 18s and 28s units, and released in the cytoplasm where they remain either free or bound to the endoplasmic reticulum.

The circular DNA molecules present in the extra nucleolus like bodiesproduce large amounts of rRNA. These extra nucleoli and the rDNA disappear gradually with the progress of embryogenesis.

Chromatin: is considered as the most important part of the nucleus as it forms the basis of heredity. It contains DNA, RNA and protein in a compact form, in which most of the DNA sequences are functionally inactive and inaccessible due to binding of proteins.

Chromatin is constructed of several subunits of each 200 base pairs of DNA organised by basic proteins called histones, into a bead like structure. The DNA cannot be directly packaged into chromatin. Therefore, it has 3 levels of organisations:

(i) winding of the DNA into bead-like particle, which are invariant competant of euchromatin, heterochromatin and chromosomes.

(ii) coiling of series of beads into a helical chain to form the chromatin fibre.

(iii) finally, the level or organisation determined by the packaging ratio of the fibre itself.

CHEMISTRY OF CHROMATIN

Chromatin is a viscous substance containing DNA, RNA and PROTEINS. It consists of more proteins than DNA. The proteins are of histones and non-histones. RNA is less than 10% of the mass of DNA and is mostly in the form of nascent chains that are still associated with the DNA template.

FUNCTIONS OF THE NUCLEUS

(1) Nucleus carries the genetic material, i.e. DNA.

(2) DNA also helps in regulation of gene expression by selective import of proteins into the nucleus. Proteins responsible for genome structure and organisation are all imported into the nucleus selectively. They include histones, DNA polymerases, RNA Polymerases, transcription factors and splicing factors. These proteins are targeted to the nucleus by specific amino acid sequences called nuclear localisation signals. These signals direct their transport through the nuclear pore complex.

In the absence of ATP proteins containing nuclear localisation, signals bind to nuclear pore complex but do not pass through the pore.

Transport of proteins through nuclear pore is an energy dependent process. It requires ATP and GTP.

Some proteins move within the nucleus and some keep moving in and out of the nucleus. These proteins are shuttling proteins. Import of proteins is regulated by either phosphorylation or by binding of proteins which cover the nuclear localization signals for importing, binding and subsequent protein transport.

RNAs are transported out of the nucleus; they are transported as ribonucleoprotein complexes. One of the proteins associated with the RNA acts as a signal for transport.

(3) Nuclear matrix provides ultraskeletal structure for the nucleus and acts as a support for various reactions and domains.

Nuclear region contains a cluster of rRNA genes. They are very actively transcribed by RNA Polymerose I.

Primary transcript of the rRNA genes is a large 45s pre-RNA which contains 18s, 28s and 5-8s rRNA. External transcribed regions are present on both 5′ and 3′ ends of the pre-rRNAs and 2 internal transcribed spacers lie between the 5, 8s, 18s and 28s rRNA. Removal of ETs and ITs give rise to independent rRNA fragments. This process is called splicing and is helped by an array of mRNAs. After splicing the rRNAs along with proteins, and 5sRNA are packed in the nucleus to form Ribosomal subunits, and they are transported out of the nucleus into cytoplasm.

PEROXISOMES

Protozoans, cells of fungi, plants, liver and kidney of vertebrates possess certain membrane bound spherical bodies of 0.2 to 1.5 μm diameter in close association with endoplasmic reticulum, mitochondria or chloroplast. They are called microbodies or peroxisomes.

Peroxisomes are small, membrane enclosed organelles that contain enzymes involved in a variety of metabolic reactions, including several aspects of energy metabolism. Though peroxisomes are morphologically similar to lysosomes, they are assembled like mitochondria and Chloroplasts from proteins, which are synthesized on free ribosomes and imported into peroxisomes as completed polypeptide chains. Though peroxisomes do not contain own genomes, they are similar to mitochondria and chloroplasts in that they replicate.

Peroxisomes occur in many animal cells and in a wide range of plants. They occur in all photosynthetic plants, brown algae, fungi and protozoans.

Peroxisomes in which glyoxylate cycle occurs are often called Glyoxysomes. They are found in yeasts, neurospora, oil rich seeds etc.

FUNCTIONS

Peroxisomes contain at least 50 enzymes involved in biochemical pathways in different types of cells.

(1) Peroxisomes are originally defined as organelles which carry out oxidation reactions leading to the production of hydrogen peroxide.

(2) As H_2O_2 is harmful to the cell they also contain catalase and enzymes which decompose hydrogen peroxide either by converting it to water or by using it to oxidise another organic compound.

(3) They are thus involved in oxidation of a variety of compounds,viz., uric acids, amino acids, and fatty acids.

(4) Oxidation of fatty acids is of great improtance, as it provides a major source of metabolic energy.

(5) In animal cells, fatty acids are oxidised only in peroxisomes.

(6) Peroxisomes are also involved in lipid biosynthesis. In animal cells cholesterol and dolichol are synthesised in peroxisomes.

(7) In liver, peroxisomes are involved in the synthesis of bile acids, which are derived from cholesterol.

(8) Peroxisomes contain enzymes required for the synthesis of plasma-logens, a family of phospholipids. Plasmalogens are important tissue components in heart and brain.

(9) Peroxisomes play a two way important role in plant cells. In seeds they are responsible for conversion of stored fatty acids to carbohydrates, which provide energy and raw materials for growth of the germinating plant. It occurs through a series of reactions forming the glyoxylate cycle. Thus peroxisomes are also called Glyoxysomes.

(10) In leaves they are involved in photorespiration which serves to metabolise a side product formed during photosynthesis. During photosynthesis CO_2 is converted into carbohydrates by calvin cycle. Through successive steps Glycine is transported to mitochondria, where it is converted to serine. Serine is sent back to peroxisome, where it is converted to glycerate which again enters the calvin cycle in the chloroplast. Thus it helps in utilising most of the carbon in the glycolate.

PEROXISOME ASSEMBLY

Peroxysomal proteins are synthesised on free ribosomes in the cytosol and are imported to the peroxisomes as complete peptide chains. Two signals are involved in targeting proteins into peroxisomes. One is the simple sequence of 3 amino acids ser-lys-leu, and the second one is a nine amino acid sequence. However, the import mechanism is not well understood.

CHLOROPLAST

Chloroplasts were first described by Nehemiah Grew and Antony Van Leeuwenhock in the 17th century. Dutrochet in 1837 observed that chlorophyll is essential for O_2 evolution from plants. Meyer first described the structure of chloroplast in 1883. Chloroplasts belong to to the group of plastids, a term coined by Schimper in 1883.

Chloroplasts are characterised by the presence of green chlorophyll pigments. There are several types of plastids, which are classified as per presence or absence of pigments, viz., proplastid, amyloplast, leucoplast, elioplast, chloro-amyloplast, chromoplast, chloroplast, pheoplast, rhodoplast, etc.

Structure: Chloroplast is in many ways similar to mitochondria. Both of these are involved in generation of metabolic energy, contain their own genome and perform various critical functions.

Plant chloroplasts are large organelles (5-10 μm long) bounded by a double membrane-chloroplast envelope. Besides these two, chloroplast has a third inner membrane system-thylakoids membrane; this forms a network of flattened discs called thylakoids. Thylakoids are arranged in stacks called grana. These three membranes divide chloroplasts into three different compartments (Fig. 3.9 and Fig. 3.9a).

(1) Intermembrane space between two membranes of chloroplasts envelope.

(2) Stroma which lies inside the envelope but outside the grana.

(3) Thylakoid lumen (Fig. 3.10).

Chloroplasts are rich in proteins, lipids, carbohydrates, nucleic acids, minerals and pigments. They also contain phospholipids such as phosphotidyl glycerol, phosphatidic acid, inositol and ethanol amine. DNA and RNA are also associated with chloroplasts. As chloroplasts synthesise starch, they store carbohydrates. Haeme and Chlorophyll are the major pigments in chloroplats, but chlorophyll-A is more prominent. Several elements in free or bound form like copper, iron, magnesium and manganese are present in this organelle.

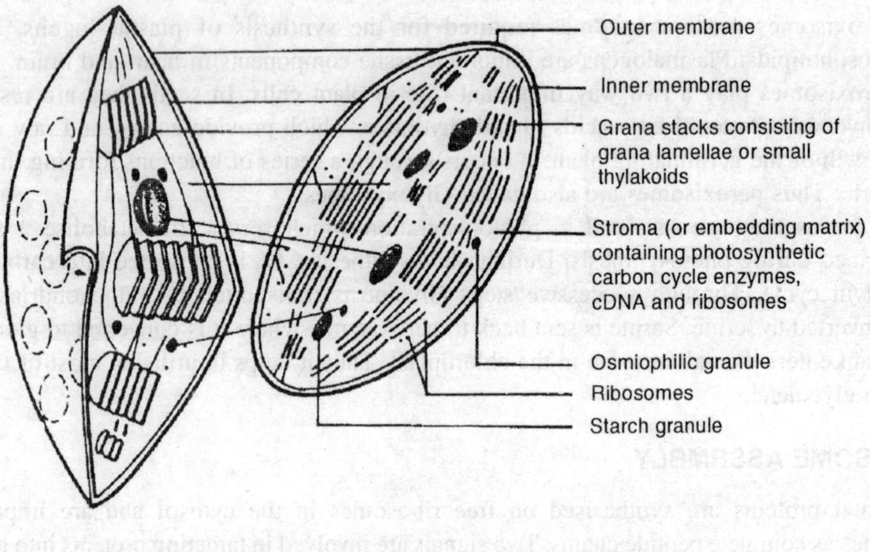

Fig. 3.9. Internal structure of the chloroplast

Chloroplast DNA (ct DNA) is present in the stroma and is capable of coding about 150 proteins which are necessary for photosynthesis and the synthesis of carbohydrates and lipids. As chloroplast DNA alone is not capable of synthesising all proteins, it is partly controlled by nuclear DNA. There is a close correlation between ct DNA diversity and speciation in higher plants.

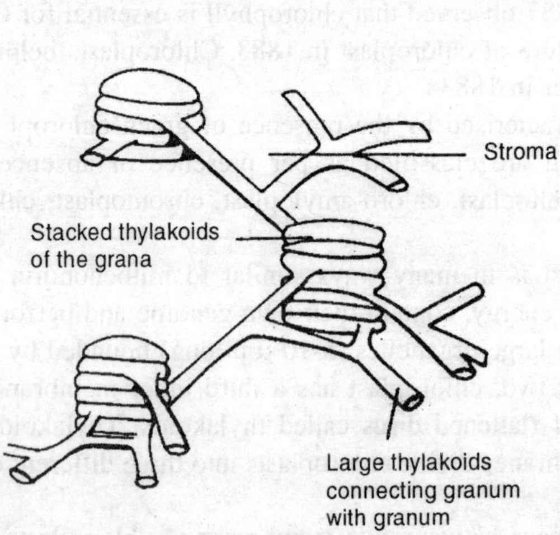

Fig. 3.9a. Internal structure of chloroplast showing arrangement of thylakoids in the stroma. The grana are interconnected

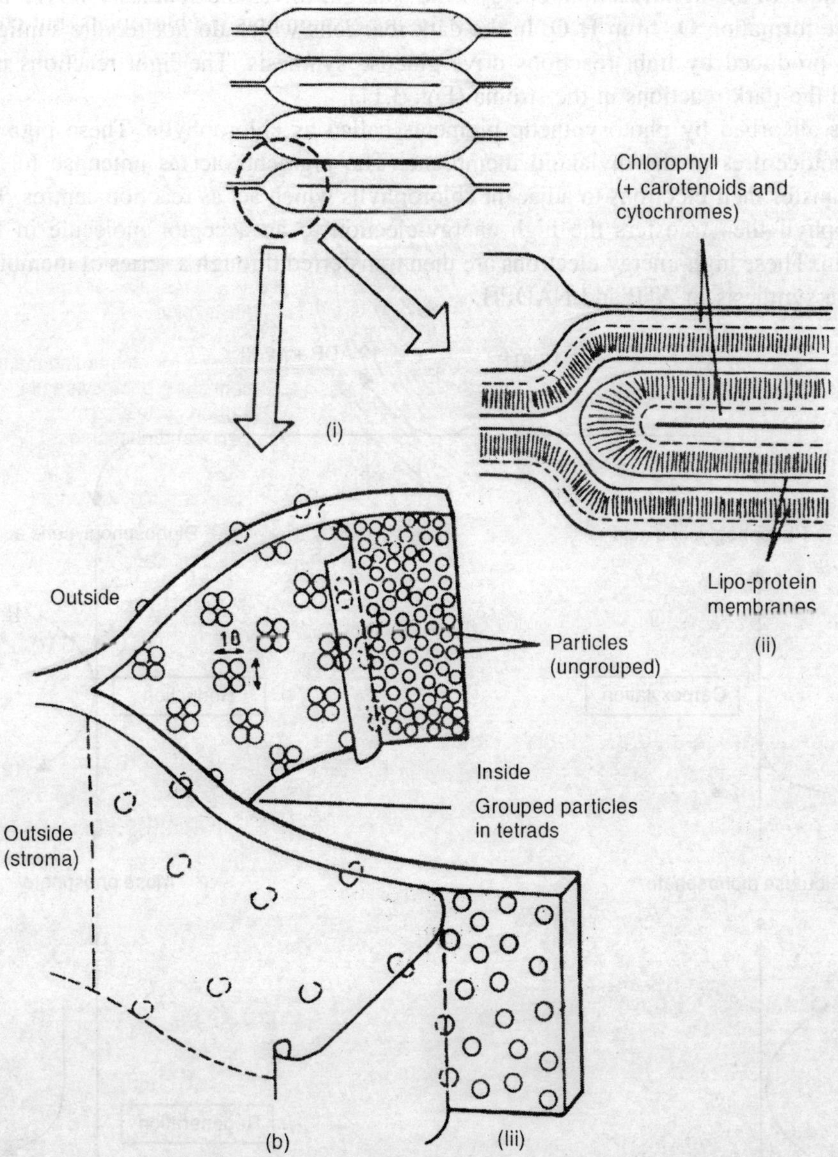

Fig. 3.10. A schematic presentation of the inner structure of thylakoids of granum: (i) arrangement of thylakoids in a stack, (ii) arrangement of chlorophyll molecules and other pigments in relation to lipoprotein membrane; (iii) thylakoid membrane showing arrangement of particles on and inside of the membrane

FUNCTIONS OF CHLOROPLASTS

1. Photosynthesis: During photosynthesis energy from sunlight is utilised to drive the synthesis of glucose from CO_2 and H_2O. Photosynthesis takes place in two distinct stages a) The light reaction b)

the dark reaction. In the light reaction, energy from sunlight drives the synthesis of ATP and NADPH coupled to the formation O_2 from H_2O. In the dark reactions which do not require sunlight, the ATP and NADPH produced by light reactions drive glucose synthesis. The light reactions occur in the thylakoid and the dark reactions in the stroma (Fig. 3.11).

Sunlight is absorbed by photosynthetic pigments called as chlorophylls. These pigments are arranged as photocentres in the thylakoid membrane. The pigments act as antennae to absorb light energy and transfer their electrons to adjacent chlorophylls which act as reaction centres. The reaction centre chlorophyll then transfers the high energy-election to an acceptor molecule in the electron transport chain. These high-energy electrons are then transferred through a series of membrane carriers coupled to the synthesis of ATP and NADPH.

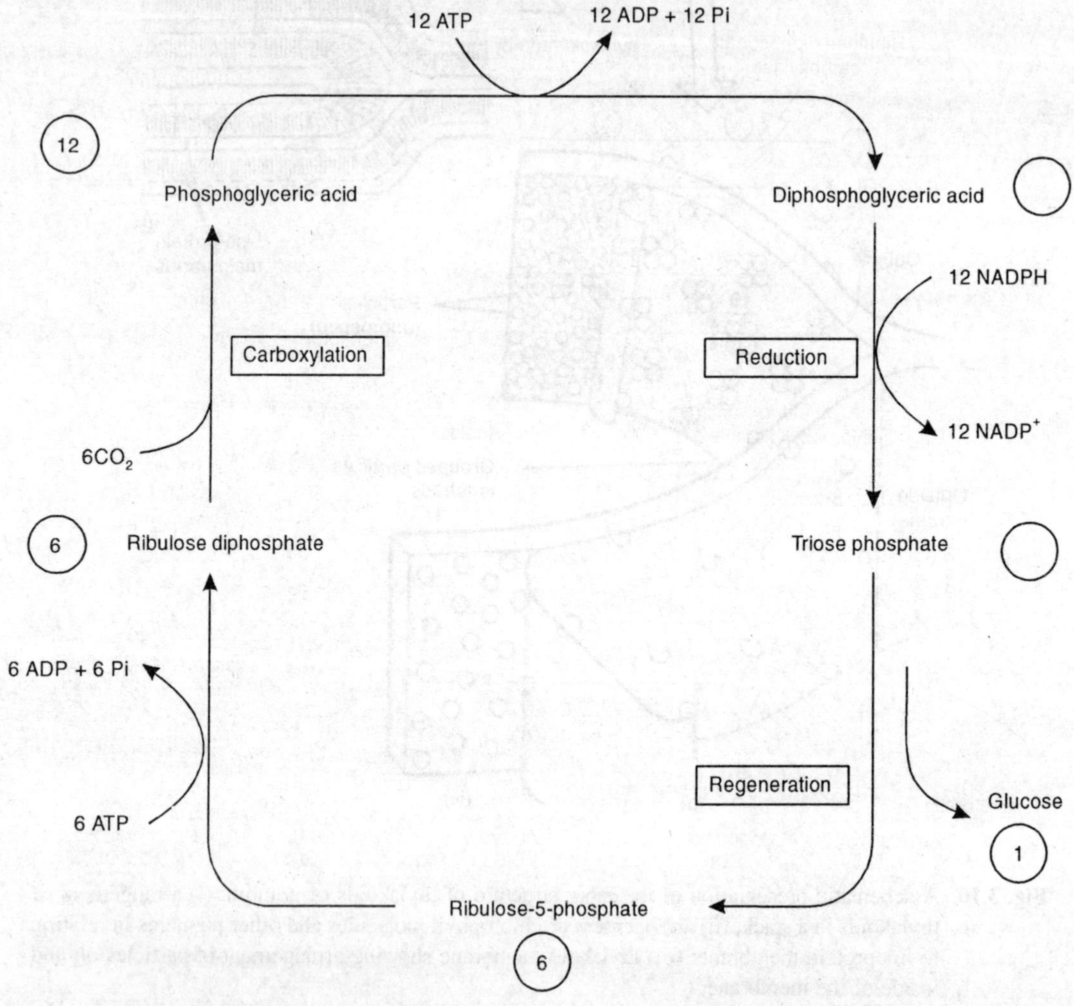

Fig. 3.11. Summary of Calvin cycle to show utilisation of ATP and NADPH generated during the light phase of photosynthesis

The proteins involved in the light reactions of photosynthesis in plants are organised into 5 complexes in the thylakoid membrane. Two of these complexes are photosystems (PS I and II) in which light is absorbed and transferred to reaction centre chlorophylls. High energy electrons are then transferred through a series of carriers in both photosystems and in a third complex, the cytochrome of complex. As in mitochondria, these electron transfer are coupled to the transfer of protons into the thylakoid lumen, thereby establishing a proton gradient across the thylakoid membrane. The energy stored in this proton gradient is then harvested by a fourth protein complex in the thylakoid membrane, the ATP synthase. It couples the flow of protons back across the membrane to AP synthesis. The energy derived from sunlight during photosynthesis is not only converted to ATP but is also used to synthesise NADPH required for the subsequent conversion of CO_2 to carbohydrate the subsequent conversion of CO_2 to carbohydrate. One photo-system is used to synthesise ATP and the others for NADPH synthesis. (PSI for NADPH and PS II for ATP).

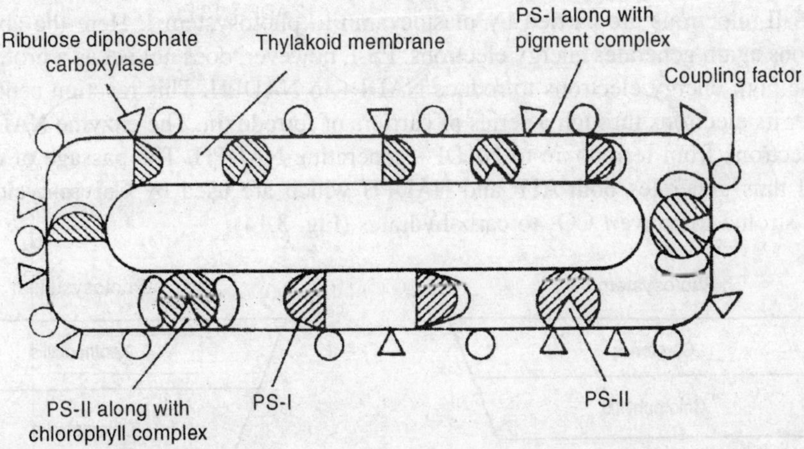

Fig. 3.12. Showing components of thylakoid membrane engaged in energy production

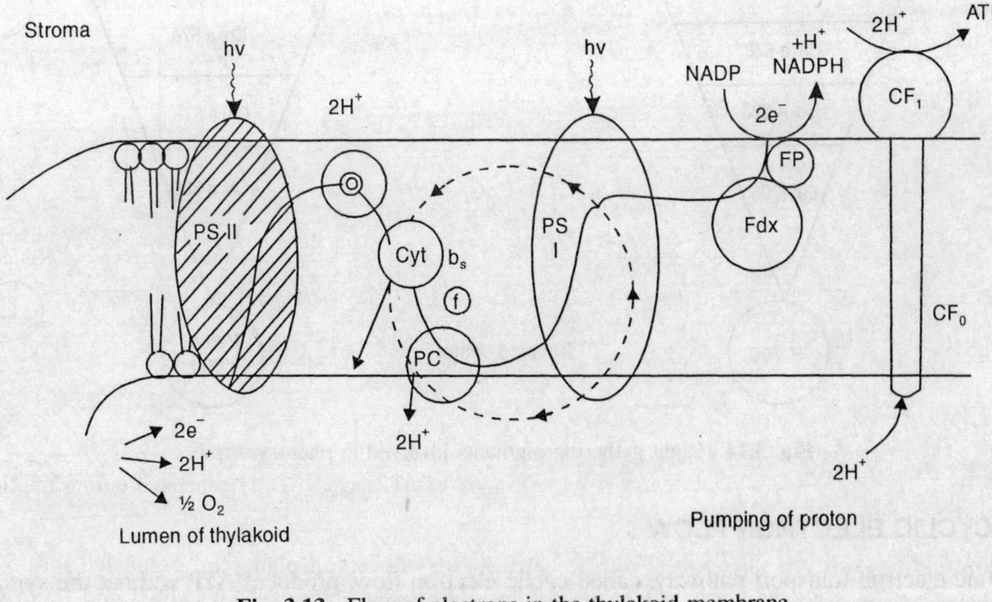

Fig. 3.13. Flow of electrons in the thylakoid membrane

The pathway of electron flow starts as PS II which is homologous to the reaction centre described in phodospirillum. At PS II the energy derived from absorption of photons is used to split water molecules to molecule O_2 and protons. This reaction takes place within the lumen of the thylakoid (Fig. 3.12).

The release of protons from water thus establishes a proton gradient across the thylakoid membrane. The high energy electrons derived from this process are transferred through a series of carriers to plastoquinone, called ubiquinone. Ubiquinone is a lipid soluble carrier similar to coenzyme-q. Platoquinone carries electrons from PS II to cytochrome of complex, within which electrons are transferred to plastocyanin and additional protons are pumped into the thylakoid lumen. Electron transport through PS II is thus coupled to establishment of a proton gradient which drives the chemiosmotic synthesis of ATP (Fig. 3.13).

From PS-II, electrons are carried by plastocyanin to photosystem I. Here the absorption of additional photons again generates energy electrons. PS-I, however, does not act as a proton pump. Instead, it uses these high energy electrons to reduce NADP+ to NADPH. This reaction centre chlorophyll of PS I transfer its electrons through a series of carriers of ferredoxin. The enzyme NADP reductase then transfers electrons from ferredoxin to NADP+ generating NADPH. The passage of electrons through PS-I and II thus generates both ATP and NADPH which are used by Calvin-cycle enzymes in the chloroplast stroma to convert CO_2 to carbohydrates (Fig. 3.14).

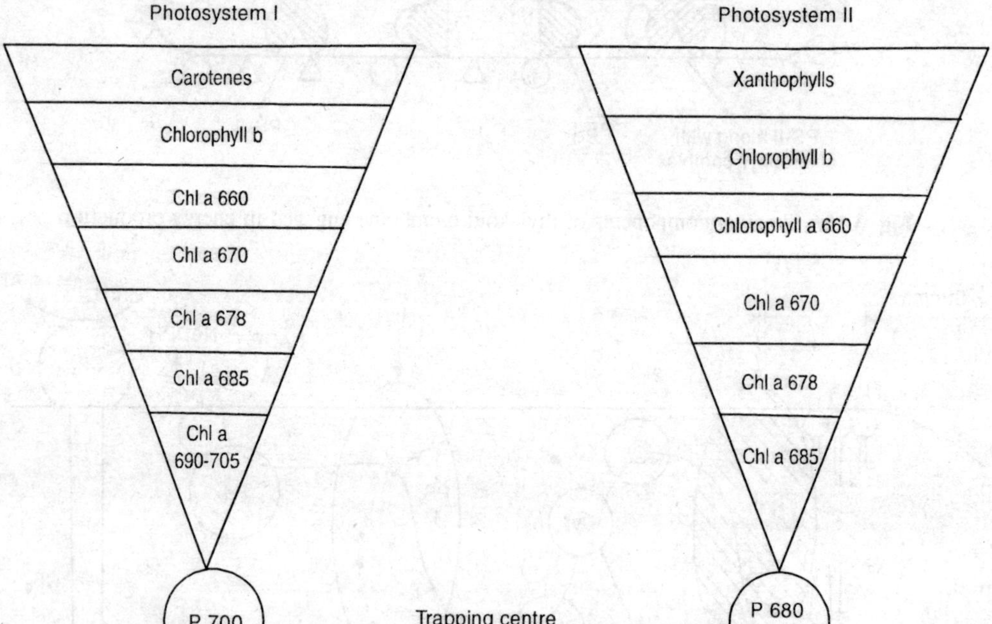

Fig. 3.14. Light gathering pigments involved in photosynthesis

THE CYCLIC ELECTRON FLOW

A second electron transport pathway, called cyclic electron flow, produces ATP without the synthesis of NADPH. It therefore supplies additional ATP for other metabolic processes. In cyclic electron flow, light energy harvested at PS I is used for ATP synthesis rather than NADPH synthesis. Instead of being transferred to NADP, high energy electrons from PS I are transferred to cytochrome of complex. Electron transfer through cytochrome of complex is then complex as PS II, to the establishment of proton gradient across the thylacoid membrane. Plastocyanin then return these electrons to PS II in a lower energy state, completing a cycle of electron transport in which light harvested at PS I is used to pump protons at the cytochrome of complex. Electron transfer from PSI can thus generate either ATP or NADPH depending on the metabolic needs of the cell (Fig. 3.15).

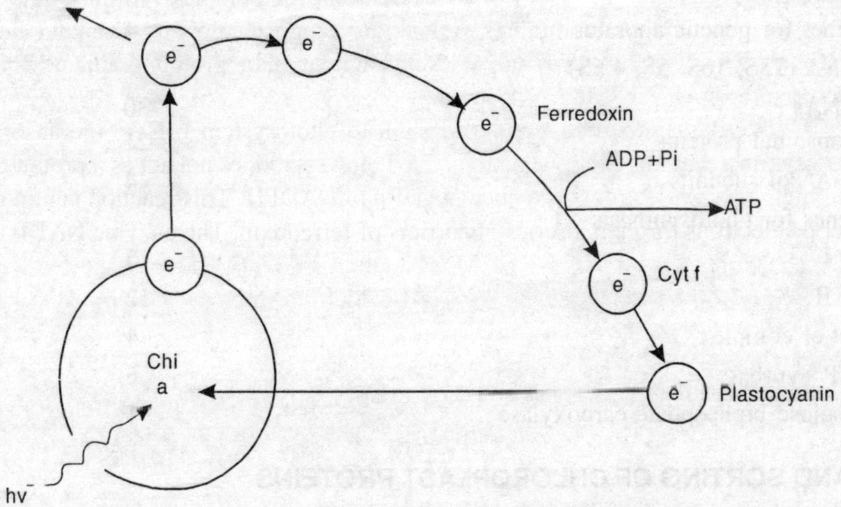

Fig. 3.15. Cyclic phosphorylation

ATP SYNTHESIS

ATP synthesis in chloroplast also occurs with the help of the enzyme ATP synthase as in mitochondria. The energy here is stored in the proton gradient across the thylacoid membrane unlike the inner mitochondrial membrane. The thylacoid is impermeable to protons but it is permeable to other ions like chlorine and magnesium. The free passage of these ions neutralises the voltage component of the proton gradient. The energy derived from photosynthesis is conserved mainly as the difference in proton concentration (pH) across the thylacoid membrane. Because the thylacoid lumen is a closed compartment, the difference in proton gradient can be quite large leading to a difference of pH about 3 units between the stroma and the lumen. For each pair of electrons transported, two protons are transported across the thylacoid membrane at PS II and two to four protons at the cytochrome complex; approximately 3 protons are needed to drive the synthesis of one ATP molecule. Each electron pair that passes through the PS I and II by noncyclic electron flow yield 1.3 and 2 ATP molecules. Cyclic electron flow has a lower yields of between 0.67 and 1.3 ATP molecules per pair of electrons.

THE CHLOROPLAST GENOME

Chloroplast contain their own genome reflecting their evolutionary origins from photosynthetic bacteria. They have multiple copies of genomes per genome. The chloroplast genome is more complex than the mitochondrial genome. It ranges from 120 to 160 kb and contains approximately 120 genes. The genome of the chloroplasts of some plants have been completely sequenced. The chloroplast genome encodes both RNA and proteins involved in the gene expression. Four rRNAs and 30tRNA species have been encoded by the genome, the rRNA include 23s, 16s, 5s and 4.5rRNA. The chloroplast genome unlike mitochondria codes for all the tRNA needed for the translation of the proteins, follow the universal genetic code.

Genes encoded by the chloroplast DNA

Function	Number of Genes
Genes for genetic appratus	4
rRNA (23S, 16S, 5S, 4.5S)	
rRNAs	30
Ribosomal proteins	21
RNA Pol subunits	4
Genes for photosynthesis	
PS I	5
PS II	12
Cyt of complex	4
ATP Synthase	6
Ribulose biphosphate carboxylase	1

IMPORT AND SORTING OF CHLOROPLAST PROTEINS

Nuclear genes code for many of the chloroplast proteins. These proteins are synthesised on cytosolic ribosomes and then imported into the chloroplast as completed polypeptides; sorting of proteins to enter the chloroplastis quite complex as the chloroplast has 3 internal separate compartments, the lumen, the inter membrane space and the stroma. Proteins are imported into the chloroplast by target sequences of N-terminal amino acids of 30-100 sequences, called transit peptides. The transit peptides direct the proteins translocation across the membranes. These are then cleaved by proteolytic cleavage. The transport of proteins across the envelope requires ATP. The proteins that are to be incorporated into the thylacoid lumen are generally targeted by two signal peptides. One is cleaved as it enters the stroma and the protein now has a second signal which helps in entry into the lumen.

OTHER PLASTIDS

Chloroplasts are only one of the members of a large family of plasbids called plastids. All plastids contain the same genome as the chloroplast. They differ in structure and function. They lack the inner membrane system (thylacoid) and the photosynthetic apparatus.

Plastids that contain pigments are generally called chromoplasts. They lack chlorophyll but contain carotenoids. They are responsible for yellow, orange and red colors of some flowers and fruits. Leucoplasts are non-pigmented plastids which store a variety of energy sources in nonphoto-synthetic tissues. Amyloplasts store starch and eleioplasts store lipids.

All plastids including chloroplasts are derived from proplastids, which are small 0.5 to 1μ diameter; they are present in organelles of rapidly dividing roots and shoot cells. They mature into plastids. Mature plastids are able to shift from one type to the other. Chlorophyll containing chloroplasts change into chloroplasts containing other pigments during ripening of fruits. The development of plastids is signalled by environmental conditions. They develop into matured ones with an intermediate vesicles, called Etioplasts, which exists in dark; when in light they develop into chloroplasts with the fusion of vesicles into thylacoid membrane systems.

SUMMARY

Cytoplasm, the term is applied to the ground substance of a living cell, that fills the interior of the cell, containing many types of organelles, each of which carries out a well-defined set of functions. This is the internal content of the cell matrix, which is the seat of most of the metabolic activities. It may, however, undergo differentiation in different types of cells, a characteristic attributed to the genes located in the chromosomes of the nucleus. The cell matrix contains highly organised array of cytoskeletal elements, the organelles and their functions are described.

EXERCISE:

1. Write in detail physico chemical nature of cytoplasm.
2. Explain the structure and functions of Golgi bodies.
3. Explain the structure and functions of endoplasmic reticulum on the basis of enzymes associated with them.
4. Explain in detail about ribosomes and their functions.
5. Write in detail about cytoskeletal structures?
6. List the organelles derived from endoplasmic reticulum and discuss briefly.
7. Mitochondria are energy converters; explain the mechanism of energy transformation by mitochondria.
8. Describe lysosomes and narrate polymorphism in lysosomes.
9. Explain the structure and functions of nucleus.
10. Explain the structure and functions of peroxisomes.
11. Describe the structure of chloroplast.
12. Narrate the functions of chloroplast.
13. What do you know about chloroplast genome?
14. Write shortnotes on:

(a) Diffusion	(h) Ribosomal RNA
(b) Cholestrol	(i) Microtubules
(c) Integral proteins	(j) Microfilaments
(d) Ion channel	(k) Secondary lysosomes
(e) Ligandgated channel	(l) Lysosomes and Diseases
(f) Vescicles	(m) Nucleolu
(g) Tight junctions	(n) Plastids

4

Chromosomes

Chromosomes are thread like structures present in the nucleus of eukaryotic cell. The term chromosome was coined by Waldeyer in 1888. In 1924, using staining techniques Fenlgen and Rossenbeck demonstrated that DNA is localised in the chromosomes.

However, the presence of chromosomes was demonstrated by the Botanist W. Hofmeister in 1849, while studying the nuclear divisions in the pollen mother cells of Tradescantia. Chromosomes are the most studied structural components of the eukaryotic cell, which are typically present in the nucleus and become clearly visible during cell division. Chromosomes play an important role in variation, heredity, mutation and evolution; as such these are considered the principal vehicles of hereditary transmission. Most of the chromosomes in a eukaryotic cell are autosomes and in higher organisms one or two chromosomes are considered as sex chromosomes which are concerned with the determination of sex. A third type of chromosomes known as supernumerary chromosomes are also found but their occurrence is quite uncommon. It is also established that the chromosomes are the carriers of genes and the total number of genes present in a haploid or single set of chromosomes in a eukaryotic cell is called genome.

A chromosome can be defined as a nuclear component endowed with a special organisation, individuality and function; it is capable of self reproduction and of maintaining its morphological and physiological properties through successive divisions.

Each thread shaped chromosome has two longitudinal subunits called chromatids. Every chromosome has a linear collection of specific genes which govern the genetic systems in the cells.

MORPHOLOGY AND STRUCTURAL ORGANISATION:

Chromosomes which are irregular thread-like structures, can be clearly observed and counted during metaphase stage of mitosis. Every chromosome has a point of attachment to the mitotic spindle, called the centromere, which determines the shape of the chromosome. In total there are four types of chromosomes (Fig. 4.1).

(1) Telocentric: In this chromosome the centromere is located on one end.

(2) Acrocentric: The centromere is located in such a way that a very short arm of chromosome is visible.

(3) Submetacentric: The centromere divides the chromosome into two unequal arms.

(4) Metacentric: The centromere divides the chromosome into two equal arms.

Morphology of an autosome is described, as they are most common.

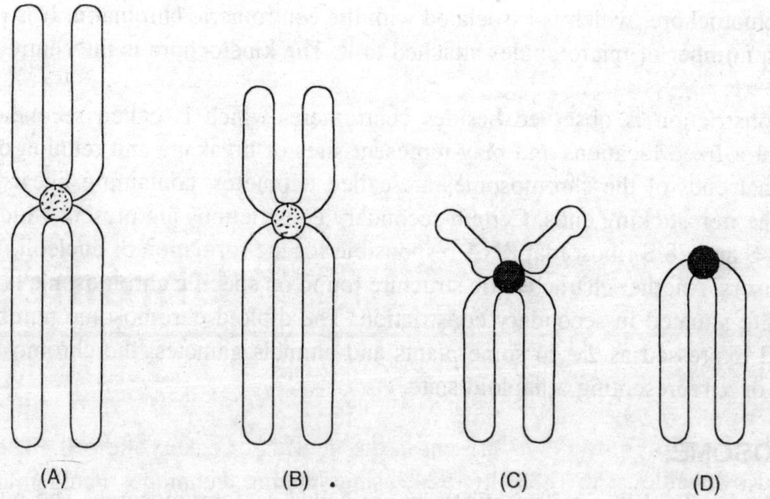

Fig. 4.1. Four types of chromosomes: A – metacentric, B – submetacentric, C – acrocentric, D – telocentric.

Every chromosome can be longitudinally divided into two chromatids, each containing a coiled chromonema. The chromonema is composed of either a single or multiple strands of nucleoprotein material rich in DNA and histones. The chromonema is surrounded by chromatin material. The chromatids are held together through a point, called centromere (Fig. 4.2) which is a point of attachment for spindle fibres during cell division. Each chromatid can be further split longitudinally in two halves, each of which is called chromonema. Thus the chromosome is a tetrad structure, every single structure consists of a multiple of chromonema fibres.

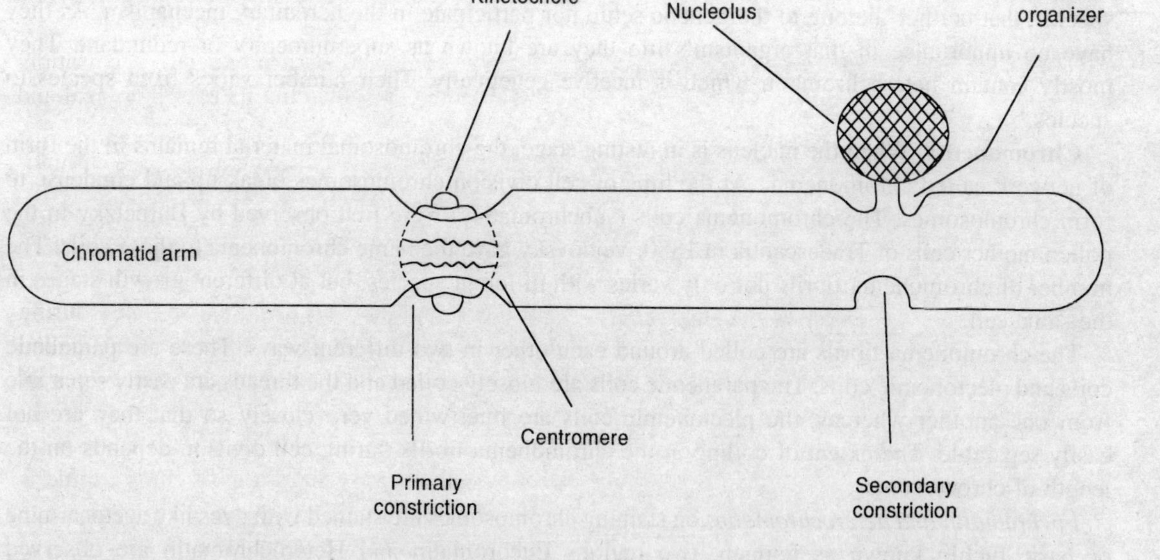

Fig. 4.2. Generalised structure of a eukaryotic chromosome showing the location of centromere and nucleolus.

Centromere is a distinct part of the chromosome morphology, which is considered as a point of constriction where the fibres of the spindle are attached during the metaphase stage. There is yet another called kinetochore, which is associated with the centromeric chromatin. It is proteinaceous in nature and has a number of microtubules attached to it. The kinetochore is the centre for assembly of microtubules.

One more constriction is observed besides centromere, which is called secondary constriction, which appears at a fixed locations and may represent sites of breakage and reunion of the chromatin material. Terminal ends of the chromosomes are called telomeres, containing linear DNA molecule. Telomeres are the non sticking ends. Certain secondary constrictions are present which contain genes that code for 18 S and 28 S ribosomal RNA responsible for the formation of nucleoli. These are called nucleolar organisers. Another characteristic structure found on specific chromosome is a spherical one, known as satellite situated in secondary constriction. The diploid chromosome number in a somatic cell is fixed and expressed as $2x$. In some plants and animals gametes, the chromosomal number is reduced to half or x representing a haploid state.

SEX CHROMOSOME:

The chromosome involved in sex-determination, are called sex chromosome. The male sex chromosome is labelled as Y and the female sex chromosome as X. In higher animals there are two different sex-determining mechanisms, these are XY and XO. Best example is Drosophila melanogaster to cite X and Y mechanism. In diploid cells of males there are three pairs of autosomes and the fourth pair consists of one X and one Y-chromosome which can be easily distinguished by their shape and size (Fig. 4.3). In female drosophila cell, in addition to three pairs of autosomes, there is a fourth pair of X-chromosomes. The XO mechanism of sex determination is found in some insects such as bugs and grasshoppers. A female grasshopper has 24 chromosomes out of which 22 are autosomes and the remaining are XX type. In males only 23 chromosomes are visible, among which 22 are autosomes and the remaining X has no partner (22 – X), hence XO mechanism

Supernumerary chromosomes: There are some extra chromosomes in many plants and animal species, that neither belong to the genetic setup nor participate in the hereditary mechanism. As they have no importance in that organism's life they are known as supernumerary or redundant. They mostly contain heterochromatin which is inactive genetically. Their number varies from species to species.

Chromonema: When the nucleus is in resting stage, the chromosomal material remains in the form of network called chromonema. At the time of cell division chromosomes break up and condense to form chromosomes. The chromonema coils (subchromatids) were first observed by Barnetzky in the pollen mother cells of Tradescantia in 1880. Vejdovsky gave the name chromonema to these coils. The number of chromonema fibrils not only varies with different species, but at different growth stages in the same cell.

The chromonema fibrils are coiled around each other in two different ways. These are paranemic coils and plectonemic coils. The paranemic coils are loosely coiled and the threads are easily separable from one another whereas, the plectonemic coils are intertwined very closely so that they are not easily separable. The extent of coiling in the chromonema fibrils during cell division depends on the length of chromosomes.

Euchromatin and heterochromatin: on staining chromosomes are stained with dyes like acetocarmine or basic fuchin known as feulgen, two regions Euchromatin and Heterochromatin are observed (Fig. 4.4). The euchromatin stains positively with the DNA specific stains indicating a concentration of DNA. This region is genetically active and stains lightly. The euchromatin regions are supposed to represent areas of less condensation. Heterochromatin stains more deeply than euchromatin and represents highly condensed regions on the chromosomes. In the interphase nucleus, the heterochromatic regions form condensed structures called chromocentres or false nucleoli.

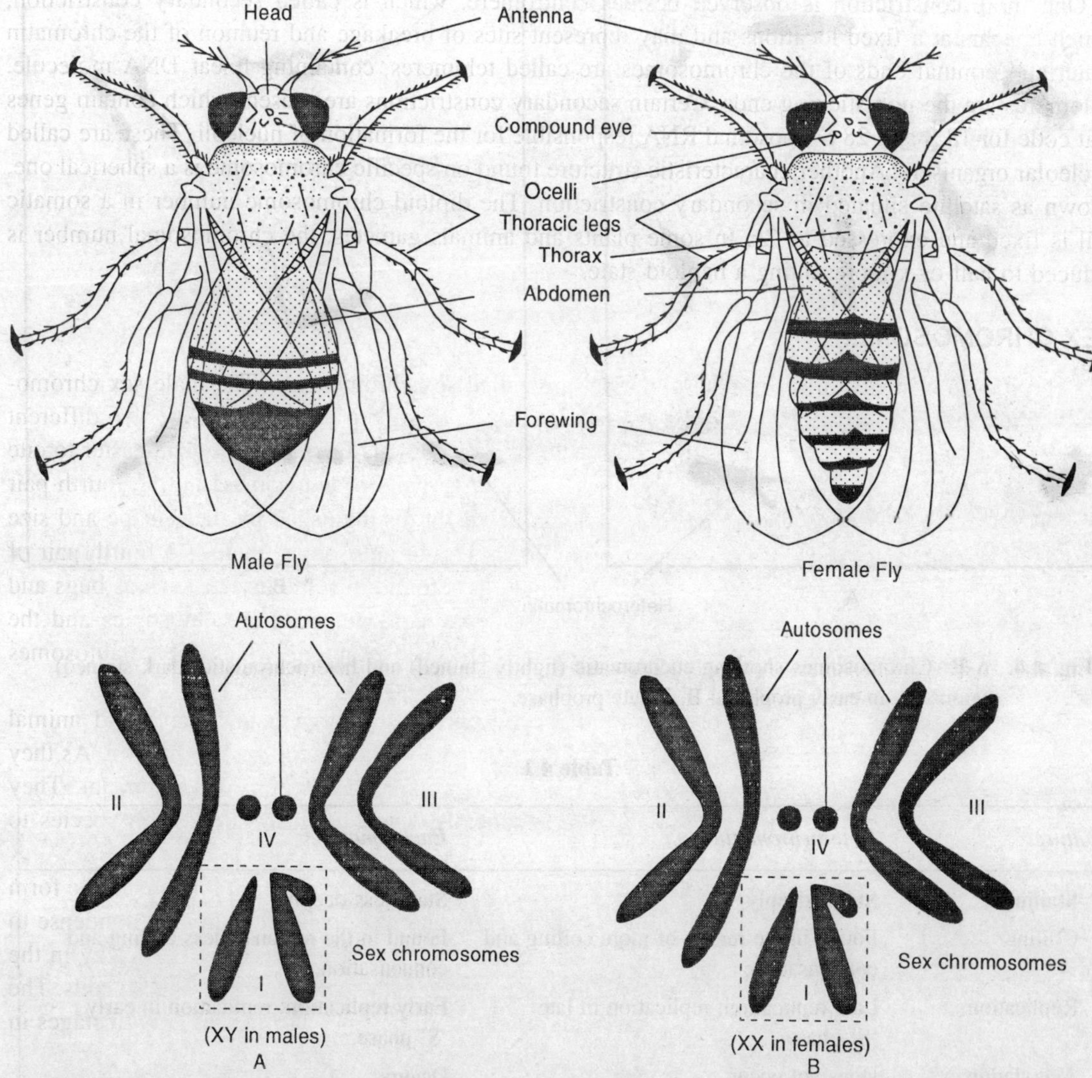

Fig. 4.3 A-B: Male and female fruit flies (Drosophila melanogaster) a diploid chromosomes complement (chromosome pair numbered chromosomes which is XY in male and XX in females and rest in II, III and IV are autosomes).

Table 4.1 summarises the differences between heterochromatin and euchromatin.

Chemical Composition: The main chemical constituents of chromosomes, include DNA, RNA proteins and lipids. Additionally metallic ions and polysaccharised are also present. DNA + Proteins constitute 90% of the chromosome. The remaining 10% is composed of residual chromosomes. Of the 90% DNA + proteins 45% is DNA and basic protein histones and protamines constitute 55%. The residual chromosome has 12 to 14% RNA, 2 to 3% DNA and 83 to 86% acidic proteins like tryptophan and tyrosine. Metallic ions include those of calcium, iron and magnesium.

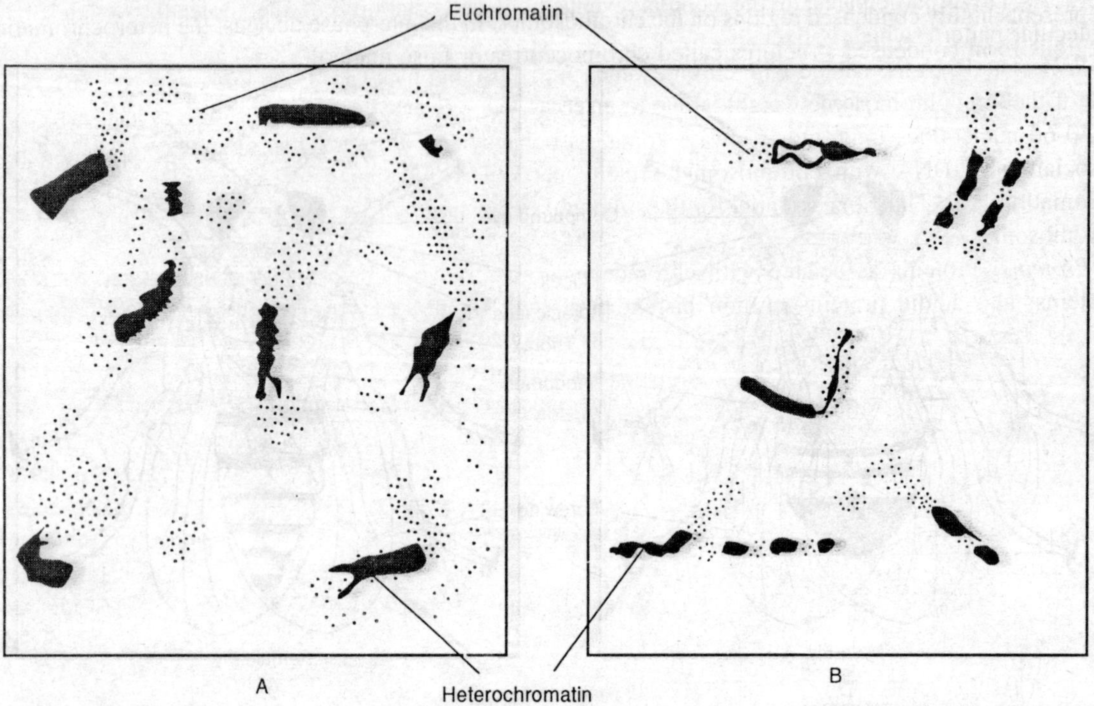

Fig. 4.4. A-B: Chromosomes showing euchromatic (lightly stained) and heterochromatic (dark stained) regions. A. in early prophase; B. in late prophase.

Table 4.1

Features	*Heterochromatin*	*Euchromatin*
1. Staining	Stains deeply	Stain less deeply.
2. Coiling	Found in the region of more coiling and condensation.	Found in the region of less coiling and condensation.
3. Replication	Late replication; replication in late 'S' phase.	Early replication; replication in early 'S' phase.
4. Acetylation	Does not occur	Occurs.
5. Effect of temperature	Pronounced	Not affected easily.
6. Genetic effect	Inert genetically but few genes are present.	Genetically active; more number of genes are present.
7. Cross over frequency	Less	Frequent.

There is convincing evidence to prove that the hereditary units, called genes are composed of DNA. The amount of DNA in all the cells of a tissue is constant and there is a definite correlation between the amount of DNA and the number of chromosome sets. DNA is very stable molecule, and once the DNA of chromosome is synthesised it does not break. This, however, does not indicate that DNA is inert. As DNA participates in the cellular events of a cell, stability of the molecule denotes its molecular pattern, which is fixed.

RNA: The RNA associated with chromosomes are ribosomal messenger and transfer RNA. Around 5% of the total chromosome weight is due to a special class of RNA called chromosomal RNA. This is 40-60 nucleotides long and is always found associated with chromosomes. It is essential for the association of DNA with chromosomal protein, particularly histone proteins in order to produce chromatin fibres. Thus it is essential for the structural organisation of chromatin fibres and therefore of chromosomes.

Proteins: Proteins associated with chromosomes are of two types — basic proteins or histone proteins, and acidic proteins or non histone proteins. About 80% of total chromosomal proteins

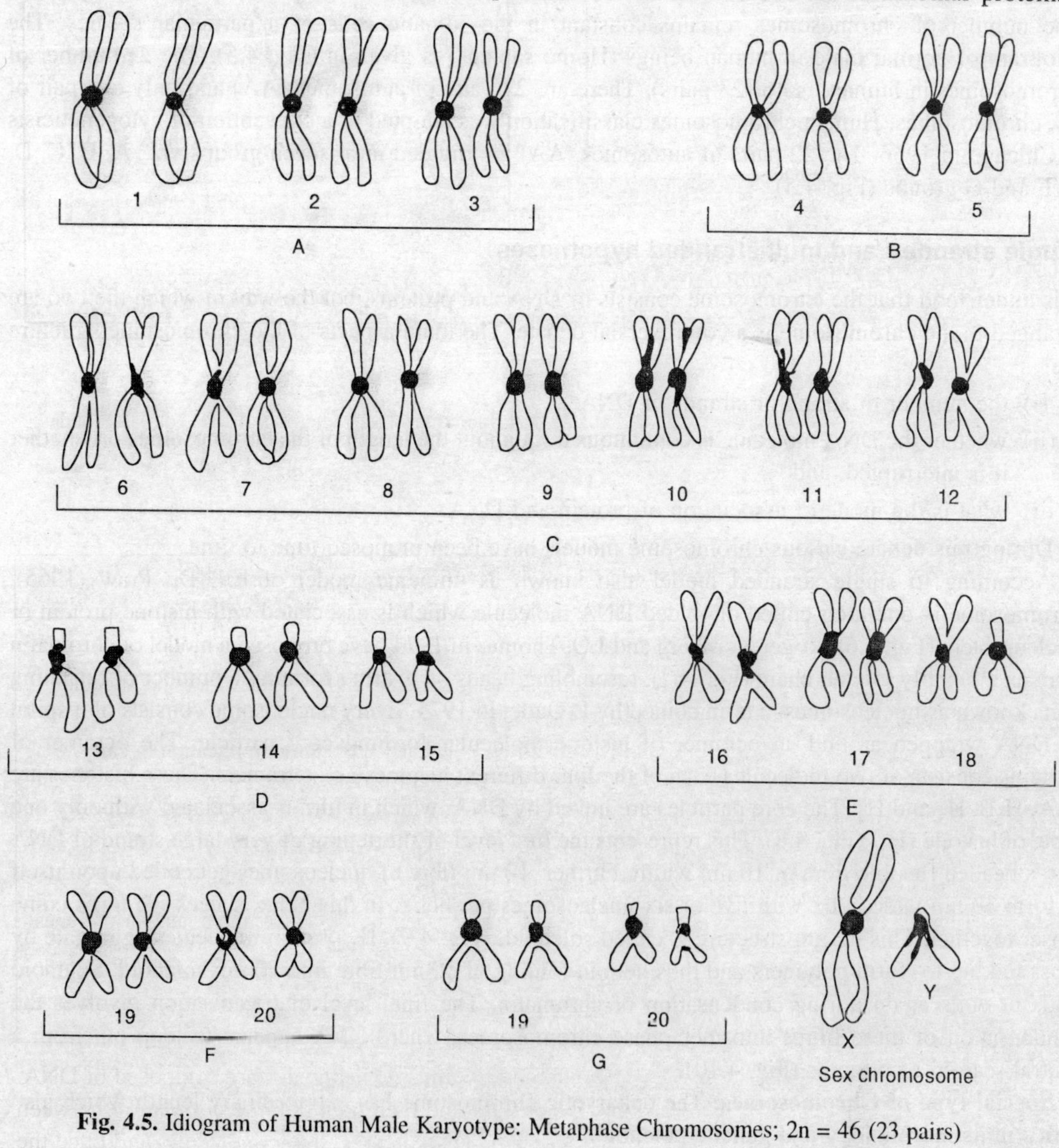

Fig. 4.5. Idiogram of Human Male Karyotype: Metaphase Chromosomes; 2n = 46 (23 pairs)

constitute histones. Its molecular weight ranges from 10,000 to 30,000 and it is completely devoid of tryptophan. In histone five different types of polypeptides are present. They are H_1, H_2a, H_2b, H_3 and H_4. Histone always consists of two copies of H_2a, H_2b, H_2 and H_4 and only one copy of H_1 polypeptide. Generally, H_1 fraction is lysine rich while H_2a and H_4 are arginine rich. H_2a, H_2b, H_3 and H_4 are involved in the structural organisation of chromatin fibres while fraction H_1 holds together the folded chromatin fibres.

Karyotype: The particular chromosome complement or set of chromosomes of a particular individual or of a species, revealed by number and morphology is called karyotype. A diagrammatic representation of the karyotype of a particular species based on number and morphology is called an idiogram. The haploid set of chromosomes containing a complete set of genes constitutes a genome. The number of chromosomes remains constant in the somatic cells of a particular species. The idiogram of normal male in human beings (Homo sapiens) is given in (Fig. 4.5). The $2x$ number of chromosomes in human, is 46 (23 pairs). There are 22 pairs of autosomes (AA) and only one pair of sex chromosomes. Human chromosomes classification was adopted in a convention of cytogeneticists at Chicago in 1966. The 22 pairs of autosomes (AA) are divided into seven groups viz., A, B, C, D, E, F and G groups (Fig. 4.5).

Single stranded and multistranded hypotheses:

It is understood that the chromosome consists of DNA and proteins, but the way in which the two are arranged in the chromosome is a controversial debate. The main aspects of the chromosome structure are:

(i) the number of strand or strands of DNA,

(ii) whether the DNA molecule is continuous throughout the length of the chromosomes or whether it is interrupted, and

(iii) what is the mode of association of protein and DNA.

During this debate various chromosome models have been proposed time to time.

According to single stranded model also known as unineme model of E.J. Du Praw (1965), chromosome is one long coiled or folded DNA molecule which is associated with histone protein or nucleoprotein (Fig. 4.6). Roger Kornberg and J.O. Thomas in 1974 have proposed a model of chromatin fibre as a flexibly jointed chain (Fig. 4.7), resembling beads on a string forming a number of repeating units known as nucleosomes, a term coined by P. Oudet in 1975. Every nucleosome consists of a spiral of DNA wrapped around an octomer of histone molecular forming core particle. The octomer of proteins consists of two molecules each of the four different histones i.e., tetramers. These histones are H_2A, H_2B, H_3 and H_4. The core particles are linked by DNA, which in turn is associated with only one type of histone (H_1) (Fig. 4.8). This represents the first level of shortening of very large strand of DNA into a beaded flexible fibre of 10 nm width. Further 10 nm fibre of nucleosomes get coiled upon itself to form 30 nm wide helix with five or six nucleosomes per helix. In this helix, successive turns come closer together. This 30 nm structure is called solenoid. (Fig. 4.9). H_1 protein molecules aggregate by cross linking to form polymers and thus help in folding of 10 nm fibre into 30 nm solenoid. Solenoid folds or coils again during condensation of chromatin. The final level of organisation involves the condensation of these fibres into metaphase chromosomes, where DNA appears to loop out from a central scafold of proteins (Fig. 4.10).

Special type of Chromosomes: The eukaryotic chromosome has extraordinary length which also varies in its morphology and genetic potential.

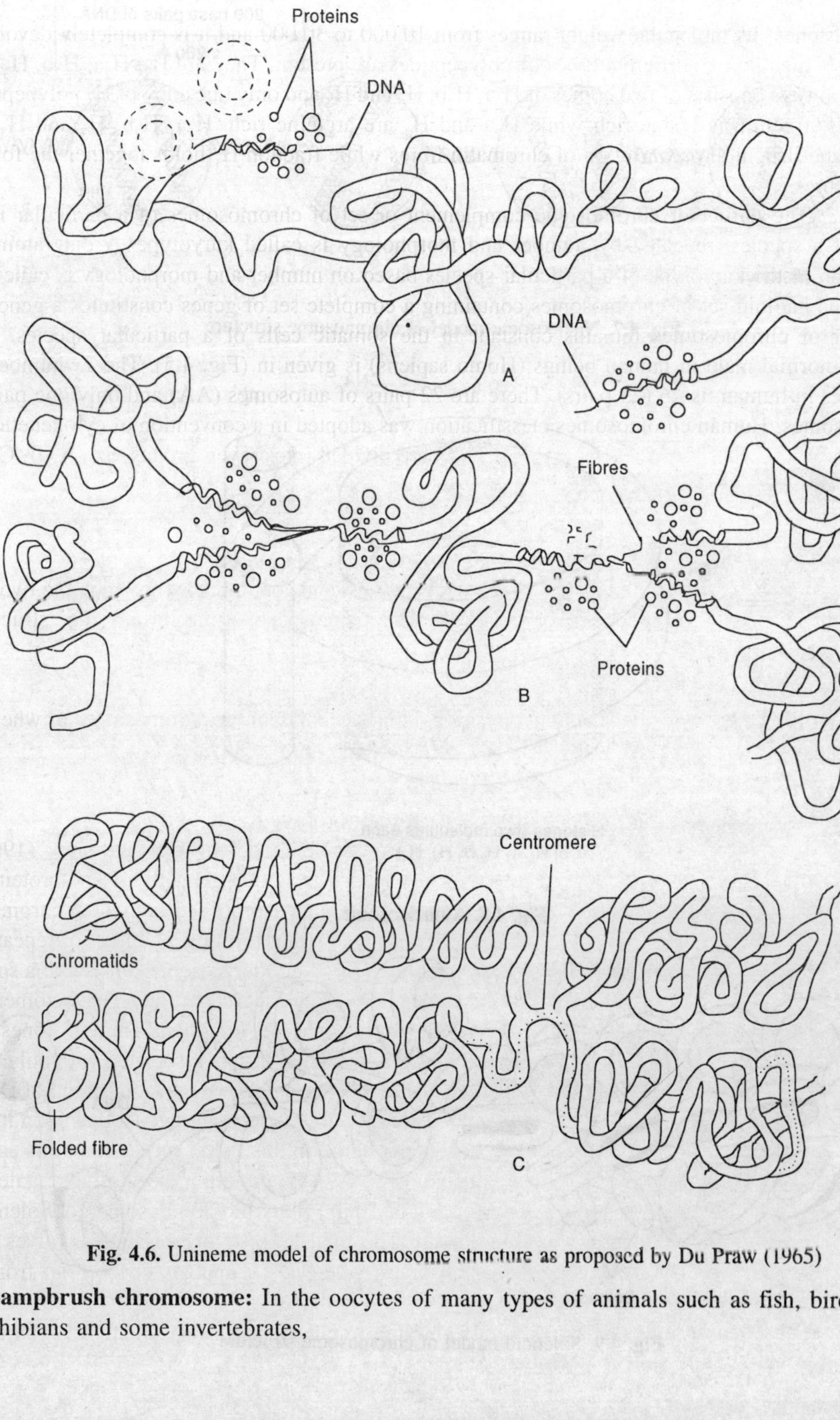

Fig. 4.6. Unineme model of chromosome structure as proposed by Du Praw (1965)

Lampbrush chromosome: In the oocytes of many types of animals such as fish, birds, reptiles, amphibians and some invertebrates,

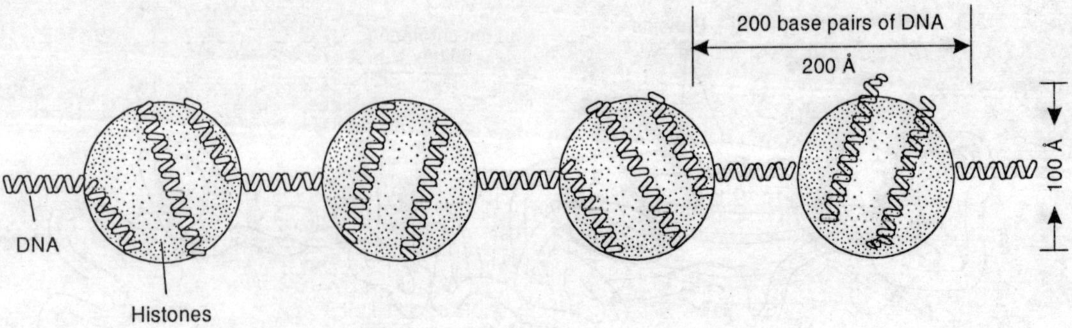

Fig. 4.7. Nucleosome model of chromosome structure

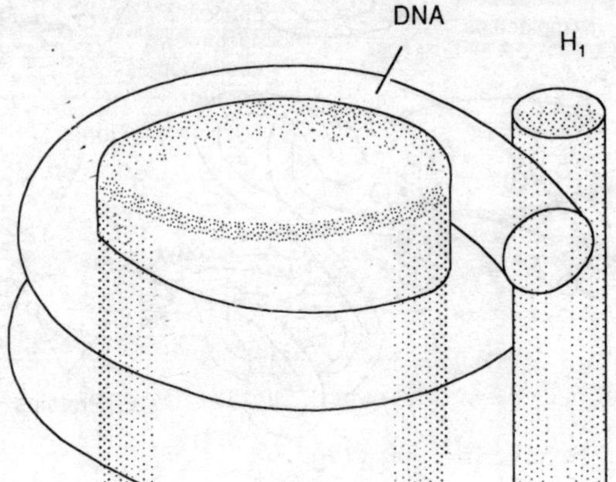

Fig. 4.8. A nucleosome.

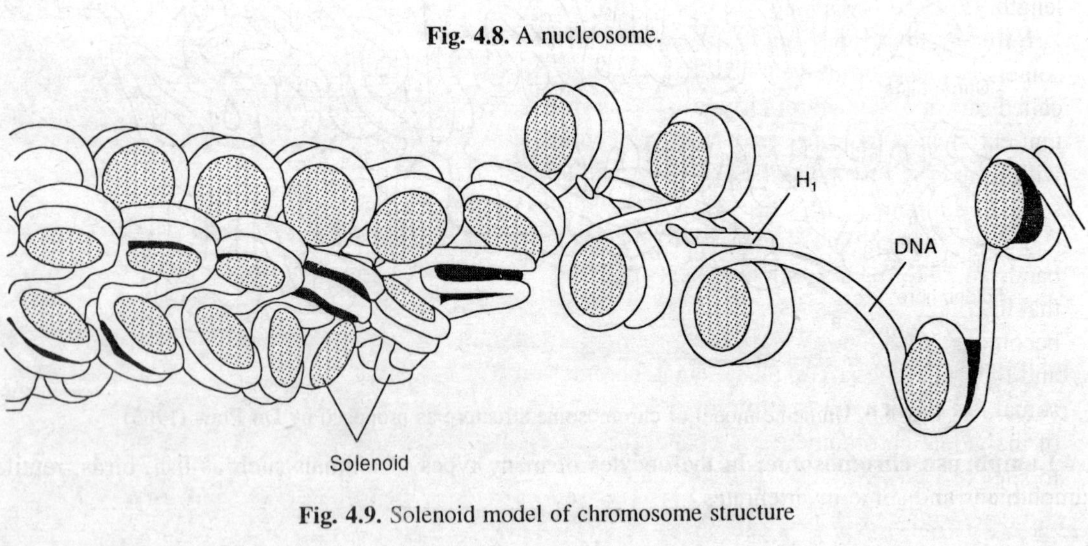

Fig. 4.9. Solenoid model of chromosome structure

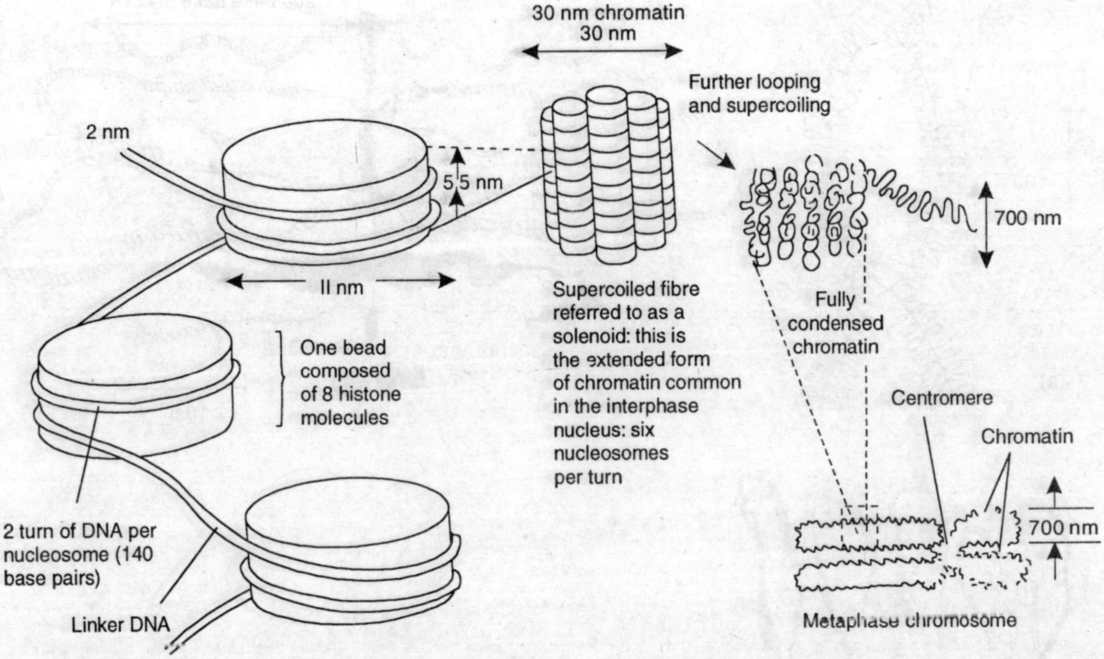

Fig. 4.10. Structure of nucleosome and its relationship to the chromosome and the DNA moelcule.

Lampbrush chromosomes have been observed, which produce a lot of RNA and grow in cytoplasmic and nuclear volume. In invertebrates these are characterised by hairy appearance. An isolated chromosome may have a maximum size of 1000 μm and consists of a row of dense granules or chromomeres held together by an extremely fine axial fibre. It has lateral loops extending from the chromomeres held together by an extremely fine axial fibre. It also has lateral loops extending from the chromomeres. As the chromosome is highly elastic, it can be stretched to about 2.5 times of its original length. (Fig. 4.11).

Salivary gland chromosome: Salivary gland chromosomes are also known as polytene chromosomes. Balbiani in the year 1881 has observed in salivary glands of midge chironomous; hence are called salivary gland chromosomes. The salivary gland chromosome is formed by about 1000 to 4000 unit chromatids (chromonemata). Every chromonemata consists of a single DNA molecule associated with protein. These polytene chromosome DNA fibres are continuous from one end of the chromosome to the other.

Painter in 1933 and Bridges in 1936 have shown that these salivary gland chromosomes possess bands and inter bands regions. They suggested that these bands are the sites of genes and it is proved that interbands regions also contain several genes. These bands of polytene chromosomes certain times become enlarged and swell; these swellings are called chromosome puffs, or Balbiani rings. Beerman and Bahr in 1954 have explained that, chromosome puffs are regions where the tightly coiled chromosomal fibres open out to form many loops. As they clarified, puffing is due to unfolding or uncoiling of individual chromomeres in a band. It is also proven that the puffs are active genes and correspond to sites of RNA synthesis. (Fig. 4.12).

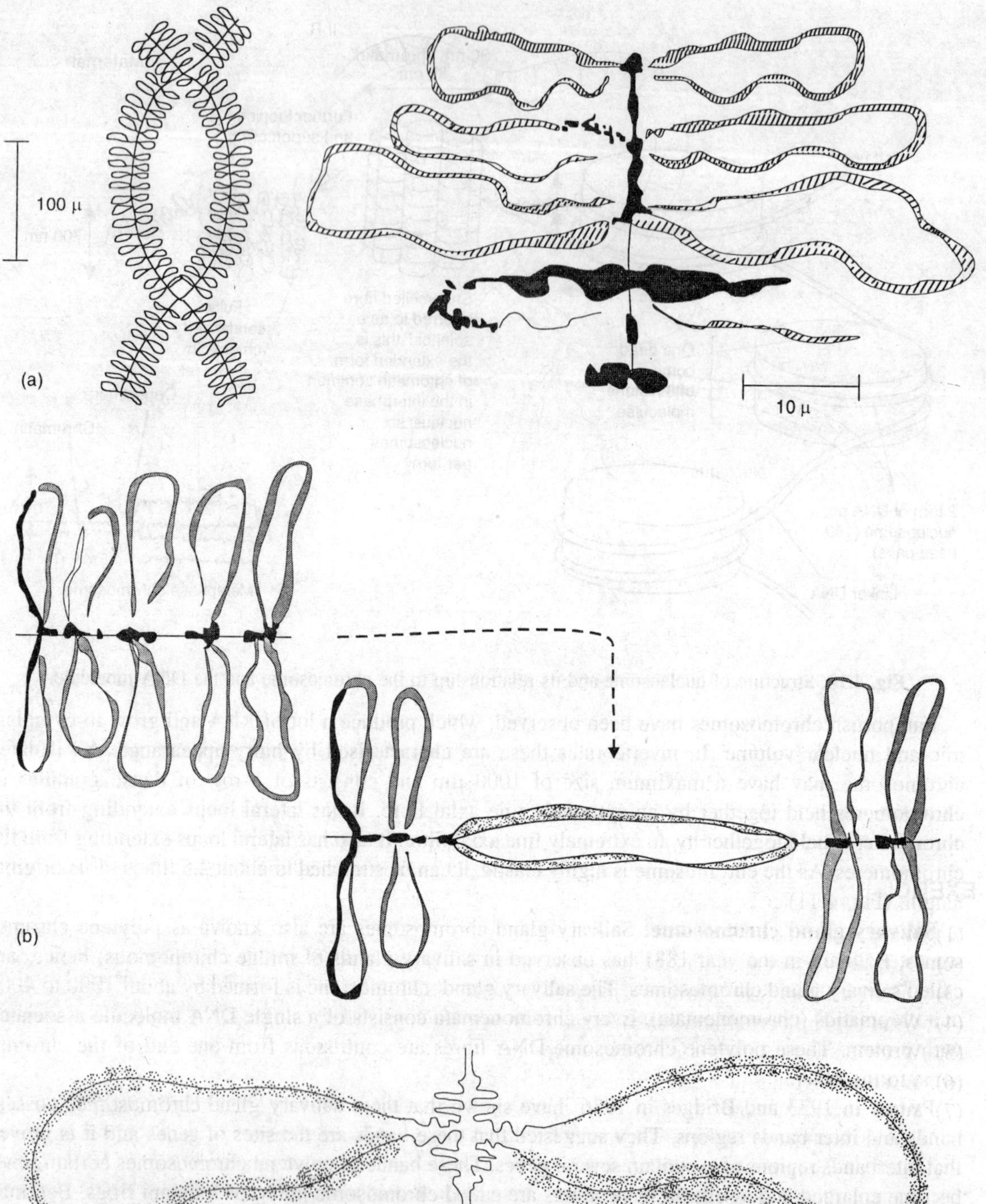

Fig. 4.11. Structure of lampbrush chromosome (a) a highly magnified view of lampbrush chromosome loops originating from the central axis, (b) showing continuity in central axis and the loops, (c) showing a gradient of nascent DNA along the DNA molecule.

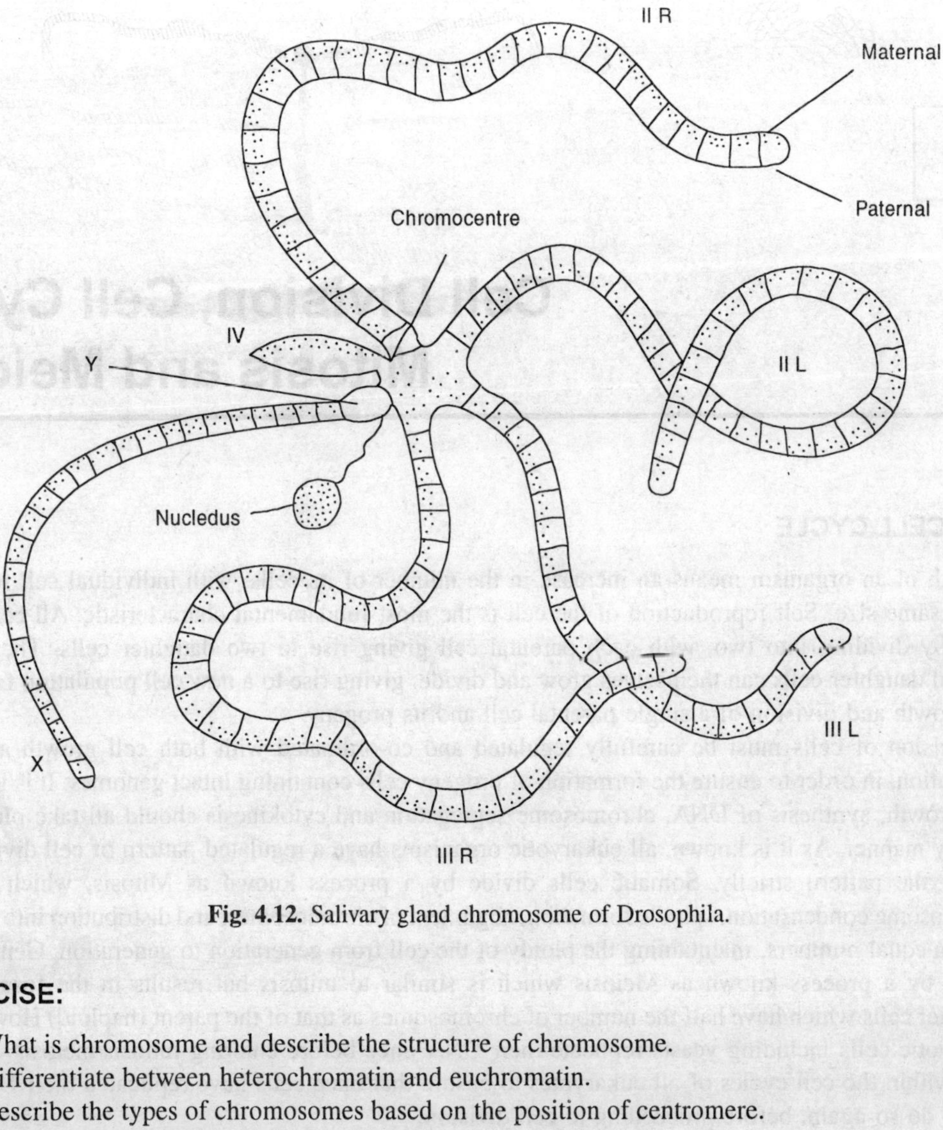

Fig. 4.12. Salivary gland chromosome of Drosophila.

EXERCISE:

(1) What is chromosome and describe the structure of chromosome.
(2) Differentiate between heterochromatin and euchromatin.
(3) Describe the types of chromosomes based on the position of centromere.
(4) What is giant chromosome? Add a note on lampbrush chromosome.
(5) Write a note on salivary gland chromosome.
(6) Write an account on single stranded and multistranded hypotheses.
(7) Write short notes on
   (i) Euchromatin and heterochromatin
   (ii) Chromonema
   (iii) Secondary constriction
   (iv) Sex chromosomes
   (v) Chemical composition of chromosome
   (vi) Karyotype
   (vii) Salivary gland chromosome.
   (viii) Lampbrush chromosome.

5

Cell Division, Cell Cycle, Mitosis and Meiosis

THE CELL CYCLE

Growth of an organism means an increase in the number of its cells, with individual cell remaining about same size. Self reproduction of the cell is the most fundamental characteristic. All cells reproduce by dividing into two, with each parental cell giving rise to two daughter cells. These newly formed daughter cells can themselves grow and divide, giving rise to a new cell population formed by the growth and division of a single parental cell and its progeny.

Division of cells must be carefully regulated and co-ordinated with both cell growth and DNA replication, in order to ensure the formation of progeny cells containing intact genomes. It is important that growth, synthesis of DNA, chromosome segregation and cytokinesis should all take place in an orderly manner. As it is known, all eukaryotic organisms have a regulated pattern of cell division and follow the pattern strictly. Somatic cells divide by a process known as Mitosis, which involves chromosome condensation, spindle formation, alignment of chromosomes, and distribution into daughter cells in equal numbers, maintaining the ploidy of the cell from generation to generation. Genetic cells divide by a process known as Meiosis which is similar to mitosis but results in the formation of daughter cells which have half the number of chromosomes as that of the parent (haploid) However, all eukaryotic cells including yeasts replicate their DNA once before entering mitosis/meiosis. Controls exist within the cell cycles of all eukaryotes to ensure that once cells have replicated their DNA they do not do so again, before proceeding to cell division.

In eukaryotic cells, progression through the cell cycle is controlled by a series of protein kinases that have been conserved from yeasts to mammals. In higher eukaryotes, this cell cycle machinery is regulated by growth factors that control cell proliferation. It is the disturbance or the defects in the cell cycle regulation that brings about abnormal proliferation of cells as in cancer cells.

EUKARYOTIC CELL CYCLE

The division cycle of most cells consists of four co-ordinated processes:

(a) Cell Growth,
(b) DNA Replication,
(c) Distribution of duplicated chromosomes to daughter cells, and
(d) Cell Division.

In bacteria, cell growth and DNA replication occurs throughout most of the cell cycle, and duplicated chromosomes are distributed to daughter cells in association with the plasma membrane. In eukaryotes, the cell cycle is more complex and consists of four discrete phases. Cell growth is a continuous process, but DNA synthesis occurs only in one phase of the cell cycle and the replicated chromosomes are then distributed to daughter nuclei by a complex series of events before cytokinesis. Progression between these stages of cell cycle is controlled by a conserved regulatory cellular signal that controls cell proliferation.

PHASES OF CELL CYCLE

Human cells in culture are typical examples of eukaryotic cells which divide every 24 hours. The cell cycle is divided hercinto two basic parts, Mitosis and interphase. Mitosis (Nuclear division) is the most dynamic stage of the cell cycle, which involves the separation of chromosomes and ends in cell division. However, both these events last only for an hour and 95% of the cell cycle is thus spent in interphase.

During interphase the chromosomes are condensed and are distributed throughout the nucleus so that the nucleus appears to be uniform. However, at the molecular level interphase is a very dynamic phase which involves growth and DNA replication both preparing the cell for the subsequent cell division in an orderly manner. The cell grows at a steady rate throughout interphase with most of the dividing cells doubling the size between subsequent cell division. The timing of DNA synthesis thus divides the cell cycle into 4 discrete phases, the 'M' Phase, the 'G' Phase, the 'S' Phase and the 'G2 Phase'.

THE 'M' PHASE

This phase of the cell cycle corresponds to mitosis, followed by cytokinesis. This phase lasts for about one hour in the human cell which takes 24 hours to complete the cell cycle. In other cell types like yeast the whole cell cycle takes only 90 minutes to pass through all the stages. Even shorter cell cycles occur in early embryonic cells, shortly after fertilisation of the egg. In this case, cell growth does not occur. Instead, they rapidly divide the egg cytoplasm into smaller cells. DNA replication occurs rapidly here and 'M' phase alternates with 'S' phase, i.e. DNA synthesis phase.

The M Phase is the phase in which mitosis occurs. This phase includes various orderly steps which end up in cell division giving rise to two daughter cells. Mitosis starts with the condensation of chromosomes followed by movement of centrioles apart from each other towards the oposite sides of the nucleus. Then microtubules grow from the two centrioles forming spindles. Some of the spindles attach themselves to kinetochores of the chromosomes followed by the alignment of these chromosomes in the metaphase plate or the equatorial plate. Then the spindles start to pull apart the chromatids, characteristic of the anophase and slowly the chromosomes move apart towards the poles. This event is followed by cytokinesis in which the cell divides into two and results in the formation of 2 daughter cells which enter the next phase of the cell cycle, the 'G' phase (Fig. 5.1).

The G1 Phase: : The M Phase is followed by a G1 phase (Gap 1) which corresponds to the interval between mitosis and initiation of DNA Synthesis. During this phase the cell is metabolically active and grows continuously but does not replicate its DNA. In a human cell the G1 phase may last about 11 hours. In early embryonic cells, as already stated, the M phase is followed by 'S' phase, there being no 'G'1 phase. This indicates that these cells do not grow, but just divide and replicate their DNA continuously till a certain point. In contrast to embryonic cells some cells in adult animals cease to divide altogether. Ex. nerve cells. Other cells divide occasionally when needed, so as to replace

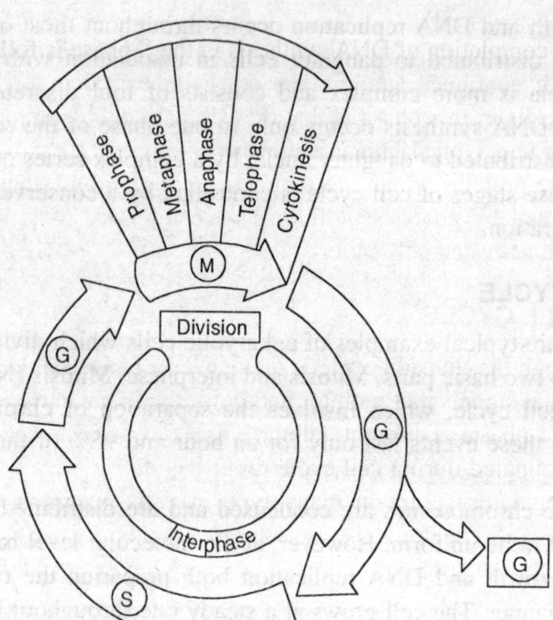

Fig. 5.1. Cell cycle

cells that have been lost because of injury or cell death. Ex. skin fibroblasts, liver cells, kidney cells and cells of the lung. These cells, after entering the G1 phase, come out of it after sufficient growth and enter a quiescent stage of the cycle called the Go phase, where they remain metabolically active but no longer proliferate unless called on to do so by appropriate extra cellular signals.

Chromosome condensation

↓

Centrioles move apart to opposite sides of the nucleus

↓

Microtubules grow from centrioles forming spindles

↓

Spindle fibres attach to kinetochores of chromosomes

↓

Chromosomes align on the metaphase plate

↓

Anaphase movement of chromatids to opposite poles

↓

Cytokinesis - cell divides

↙ ↘

Daughter cell in G1 phase | daughter cell in G1 phase

Flow chart of events occurring in M-phase.

The 'G' 2 Phase: The completion of DNA synthesis in the S phase is followed by G2 phase (Gap2) during which cell growth continues and proteins are synthesised in preparation for mitosis. The significant characters of cells in this phase is that they contain double amount of DNA i.e. (4x). This phase lasts about 4 hours in a typical human cell. The G2 phase cells are involved in active protein synthesis, especially those proteins which are required for the mitotic division. The DNA content of the cells can be determined by analysis of fluorescence intensity of individual cells in a flow of cytometer of fluorescence activated cell sorter.

REGULATION OF CELL CYCLE

Progression of cells through the division cycle is regulated by extra cellular signals from the environment, as well as by internal signals that monitor and co-ordinate the various processes that take place during different cell cycle phases. Different cellular processes such as cell growth, DNA replication and mitosis must be coordinated during cell cycle progression.

A major cell cycle regulatory point in many type of cells occurs in the late G1 and controls progression from G1 to S. This check point was defined in budding yeast (Saccharomyces cerevisiae). It was named as START (Fig. 5.2).

Once cells have passed START, they are committed to enter 'S' phase and undergo one cell division cycle. Passage through START is a high regulated event in yeast cell cycle. It is controlled by external signals, such as the availability of nutrients as well as by cell size. If nutrients are not properly available the cells arrest their cell cycle at START and enter into the Go phase rather than the S phase. Thus START represents a decision point at which the cell determines whether sufficient nutrients are available to support progression through the rest of the division cycle. Polypeptide yeast mating factors that signal yeast mating also arrest cell cycle at START. In addition to serving as a decision point for monitoring extracellular signals, START is the point to which cell growth is co-ordinated with DNA synthesis and cell division. It functions analogously to START in yeasts. However, the passage of animal cells through the cell cycle is regulated primarily by the extra cellular growth factors that signal cell proliferation, unlike yeast cells where the regulation is by avaialability of nutrients. In the presence of appropriate growth factors, cells pass the restriction point and enter S phase. Once it has passed through the restriction point, the cell is committed to proceed through 'S' phase and the rest of the cell cycle, even in the absence of further growth factor stimulation. In the absence of growth factors, cells arrest at the cycle at the restriction point and enter S phase. Once it has passed through the restriction point, the cell is committed to proceed through S phase and the rest of the cell cycle, even in the absence of further growth factor stimulation. In the absence of growth factors cells arrest at the cycle at the restriction point and enter into Go phase where the cell remains inactive for long periods Ex. skin fibroplast.

Some cells are also regulated in the G2 phase. This is seen in yeasts, where it is regulated primarily by control of transition from G2 to M. In animals, the primary example of cell cycle control in G2 is provided by oocytes. Vertebrate Oocytes remain arrested in the G2 Phase for long period of time until their progression to 'M'Phase is triggered by hormonal stimulation. Extracellular signals can thus control cell proliferation by regulating progression from the G2 to M as well as G1 to S phases of the cell cycle.

CHECK POINTS OF CELL CYCLE

Other than the controls described earlier, events that take place during different stages of the cell cycle must be coordinated with one another so that they occur in appropriate order. This coordination between different phases of the cell cycle is dependent on a system of check points and feedback controls that prevent entry into the next phase of the cell cycle until the events of the preceding phase have been completed.

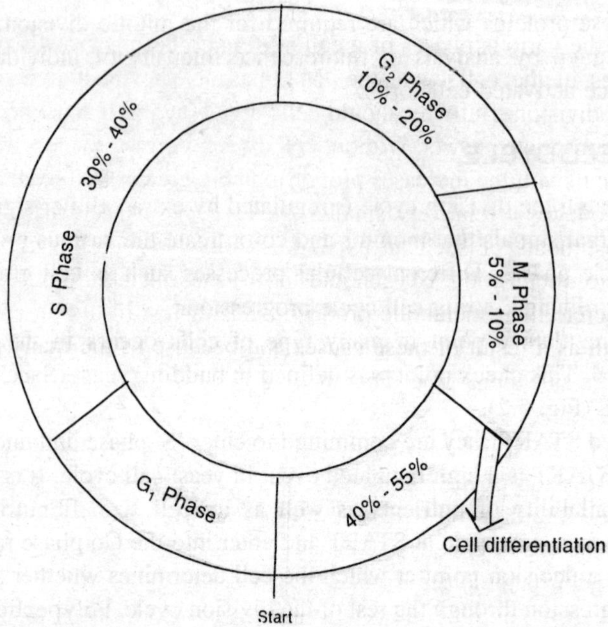

Fig. 5.2. Cell cycle: G_1 is pre-DNA synthetic phase; S phase is synthetic phase representing period of DNA synthesis; G_2 phase is post-DNA synthetic period; M phase is the mitotic phase from which some cells leave the cycle to undergo differentiation. Duration of each phase is given in %

Several cell cycle check points function to ensure that incomplete or damaged chromosomes are not replicated and passed on to daughter cells. One of the most clearly defined of these check points occurs in G2 and prevents the initiation of mitosis until DNA replication is completed. This G2 checkpoint senses unreplicated DNA and generates a signal that leads to cell cycle arrest. Operation of G2 check point prevents the initiation of 'M' phase before completion of 'S'phase, so cells remaining G2 until the Genome have been completely replicated. The cell cycle can also be arrested at the G2 check point in response to DNA damage. DNA damage also arrests the cell cycle at the G1 phase and the end of 'M' phase, which ensure damaged DNA repair, replication and also separation of chromosomes only after complete complementation has occured. Further the cell cycle is regulated by regulators like MPF (Maturation promoting factor) which is a dimer of Cd c and cyclin, a set of protein kinases. MPF is a conserved regulator of the cell cycle and is composed of two key subunits Cd c, and cyclin B. MPF activity is controlled by periodic accumulation and degradation of cyclin B during cell cycle progression. Phosphorylation of Cd c has further been provided to be the other regulatory mechanism.

In mammalian cells Cyclin B synthesis begins in the S phase. Cyclin B then accumulates and forms complexes with Cd c protein kinase. Once activated the Cd c protein kinase phosphorylates a variety of target proteins that initiate the events of M Phase. In addition, Cd c activity triggers the degradation of cyclin B , which occurs as a result of ubiquitin mediated proteolysis. The proteolytic degradation of cyclin B then inactivates Cd c leading the cell to exist mitosis, undergo cytokinesis and return to interphase. Thus the cell cycle is well regulated and plays a significant role in the life of a cell and cell division.

CELL DIVISION

As we have already seen the various phases of a cell's life, i.e., during growth and division we know that out of all the phases in the cell cycle the 'M' Phase is the most dynamic wherein the cell undergoes a programmed division. But why should cells divide at all? It must be surprising to put this question but there is an appropriate answer. Without cell division growth occurs through an increase by the square of its radius (r); its volume increases proportionately greater by the cube of this number, i.e. an increase in size (r) produces a relatively smaller increase in surface area (r) than in volume (r). Therefore, the inner constituents of such an expanding organism would have proportionately less surface area from which to obtain food, oxygen and the various metabolic necessities as well as less surface from which to secrete their metabolic products and wastes. In the absence of division, death would quickly ensure both as a result of these causes and because of the many physical stresses and accidents that could rupture as a membrane. Therefore, some form of cell division is necessary for the maintenance of a cell.

Eukaryotic cell division, as mentioned earlier, is quite complex and involves a unique mechanism. Two basic types of cell division have been identified according to the pattern of division and the behaviour of chromosomes. The division which results in the distribution of daughter chromosomes in equal numbers to each of the daughter cells and exactly the same number as in the parent cell is called Mitosis. This type of division takes place in all somatic cells. In the other type of cell division the daughter cells have only half the number of chromosomes as in the parent cell. This division is called Meiosis. It takes place in the formation of gametes or spores which are required for sexual reproduction. The first account of Mitosis was provided by A. Schneider in 1873. Walter Flemming who observed cell division in salamander and showed that nuclear division involves longitudinal splitting of the chromosomes. In 1882, he coined the term Mitosis. Cell division in prokaryotes and eukaryotes can generally be divided into three types i.e., direct cell division-amitosis, indirectcell division-mitosis and reduction division-meiosis.

MITOSIS

The cell division occurring in all somatic cells wherein the two daughter cells formed contain an equal number of chromosomes and the same number as that of the parent cell is called Mitosis. In this division the ploidy of the cell is maintained. Mitosis can be divided into four different phases. However, before entering the M phase (Mitotic), the cell is said to be in a resting stage, i.e. the interphase.

The interphase

Interphase is the period between successive cell divisions where the cell grows in size and also prepares for Mitosis. This phase includes the G1 phase, S phase and the G2 phase of the cell cycle. In all the phases the cell is not resting or inert, but is actively involved in protein synthesis and DNA replication preparing itself for Mitosis. Hence, this can be called the preparatory phase. In this phase the cytoplasm is rich with ribosomes and protein constituents. The nucleus is distinct with a nucleolus. The chromatin is rather decondensed. The time taken for mitotic division however varies from species to species (Fig. 5.3).

Prophase

It is the first stage in mitotic division. Beginning of the prophase is marked by the appearance of condensed chromosomes each of which consists of two sister chromatids. These newly replicated DNA molecules remain intertwined throughout the S and G2 phase becoming untangled during the

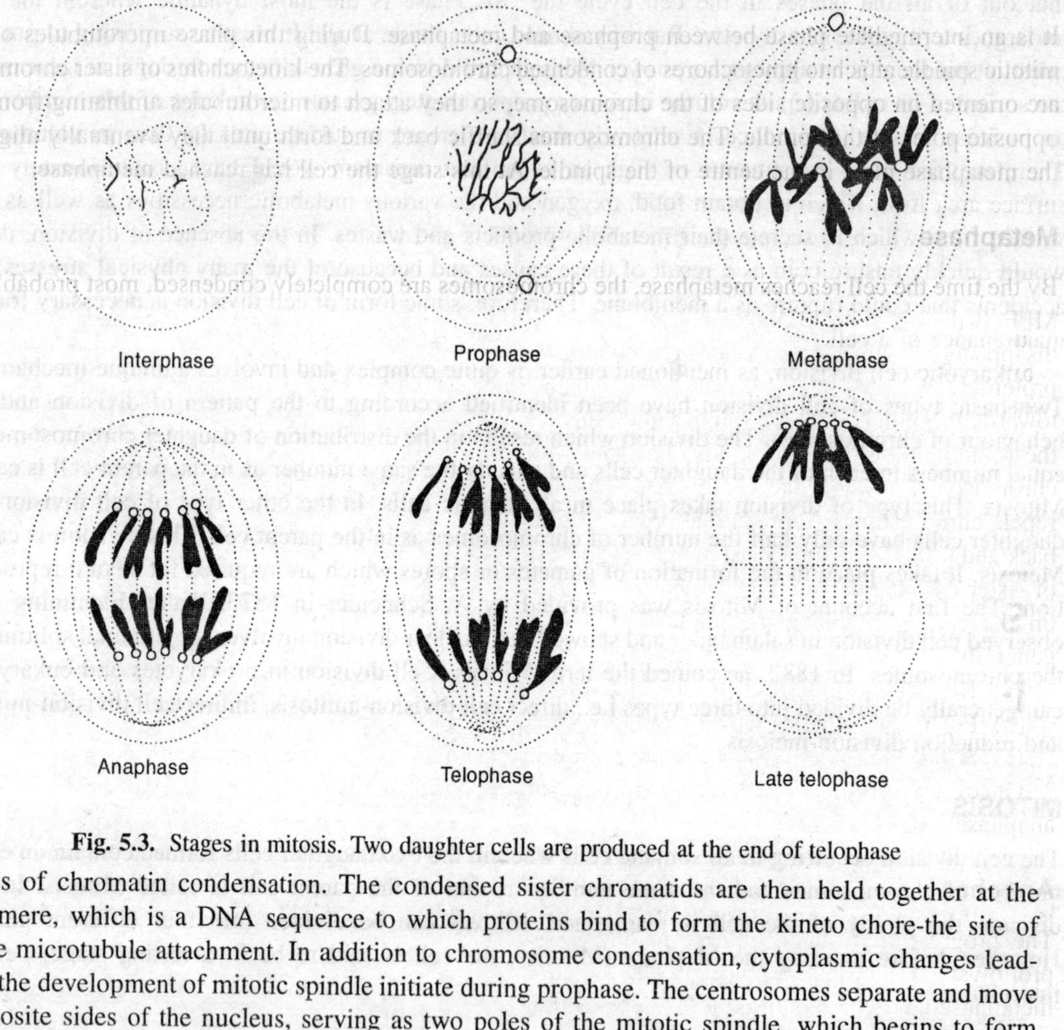

Fig. 5.3. Stages in mitosis. Two daughter cells are produced at the end of telophase

process of chromatin condensation. The condensed sister chromatids are then held together at the centromere, which is a DNA sequence to which proteins bind to form the kineto chore-the site of spindle microtubule attachment. In addition to chromosome condensation, cytoplasmic changes leading to the development of mitotic spindle initiate during prophase. The centrosomes separate and move to opposite sides of the nucleus, serving as two poles of the mitotic spindle, which begins to form during late prophase. In the end of the prophase the breakdown of nuclear envelope is seen. In most cells breakage of the nuclear envelopes marks the end of prophase of mitosis. The dissociation of nuclear lamina results from the phosphorylation of the lamines. This causes the filaments to break down into individual lamine dimerse, which is catalysed by the Cd c protein kinase. Simultaneously, the nuclear membrane also breaks up into fragments and then into vesicles.

The other major change that occurs in prophase is the chromosome condensation. The interphase chromatin, i.e. packaged into nucleosomes condenses approximately 1,000 fold to form the chromosomes. This condensation is needed to allow the chromosomes to move along the mitotic spindle without becoming tangled or broken during their distribution to daughter cells. DNA in this highly condensed stage can no longer be transcribed and so RNA synthesis stops and the nucleolus also disappears. Phosphorylation of H histone protein by Cd c protein kinase has been suggested to be involved in chromosome condensation. The condensation of chromosomes and dissociation of the nuclear membrane mark the end of prophase and cells enter prometaphase.

Prometaphase

It is an intermediate phase between prophase and metaphase. During this phase microtubules of the mitotic spindle attach to kinetochores of condensed chromosomes. The kinetochores of sister chromatids are oriented on opposite sides of the chromosome, so they attach to microtubules aminating from the opposite poles of the spindle. The chromosomes shuffle back and forth until they eventually align on the metaphase plate in the centre of the spindle. At this stage the cell has reached metaphase.

Metaphase

By the time the cell reaches metaphase, the chromosomes are completely condensed, most probably by MPF (Cd c -Cyclin B) mediated phosphorylation of H Histone protein. The nuclear membrane disappears allowing the spindle microtubules to attach to the kinetochores of the chromosomes. The proteins assembled in the kinetochore include motors that drive the movement of chromosomes towards the minus ends of spindle microtubules, which are anchored to the centrosome. Other than these motors, the centrosome also has some motor proteins bound to the centrosome which pull the chromosomes towards the centrosome. This attraction is opposed by a force called polar wind that repels chromosomes from spindle poles. Consequently, the chromosomes in the prometaphase move back and forth. Microtubules from opposite poles of the spindles eventually attach to the two kinetochores of sister chromatids and balance the forces acting upon them. Thus finally, chromosomes are aligned on the equatorial plate, the identification of metaphase. The spindle formed during the process generally contains three types of tubules.

(a) Kinetochore microtubule that attach to kinetochores
(b) Polar microtubules which overlap with one another in the centre of the cell
(c) Astral microtubules that radiate outward from the centrosome toward the cell periphery.

Once chromosomes are aligned the cells proceed into the next phase of cell division, i.e. the anaphase.

Anaphase

The progression from metaphase to anaphase results from the activation of a ubiquitin mediated protiolysis system that degrades cyclin B and thereby inactivate MPF. However, the transition from metaphase has been identified to be dependent upon a target protein X, i.e suggested to have a role in holding the two sister chromatids together at the centromere. Once this protein is degraded the chromatids are free and the movement of these chromatids to opposite poles occurs by the motor aided depolymerisation of the microtubules.

Anaphase is characterised by the movement of sister chromatids to opposite poles. Chromosome movement has been identified to proceed via two mechanisms called Anaphase A and Anaphase B. Anaphase A consists of the movement of chromosomes towards spindle poles along the kinetochore microtubules. This type of chromosome movement is thought to be driven by a kinetochore associated motor protein that translocates chromosomes along spindle microtubules in the minus end direction, towards the centrosomes. The action of these proteins is coupled to disassemble the shortening of the kinetochore microtubules. It is thought that motor activity and microtubule disassembly both contribute to chromosome movement. Anaphase B refers to the separation of spindle poles themselves. Spindle poles separation can occur by overlapping polar microtubules sliding against one another, pushing the spindle poles apart. This movement is aided by positive end directed motor proteins. Thus, separation of sister chromatids to opposite poles mediated by various proteins marks the end of the shortest phase of mitosis. Anaphase is followed by the final phase of mitosis before before cytokinesis – the Telophase.

The telophase

Telophase is marked by the reassembly of nuclei in each of the daughter cells, still to be separated, and the decondensation of chromosomes also is seen. In the telophase two nuclei form around the separated sets of daughter chromosomes. Chromosome decondensation, reassembly of nuclear envelope appear to be signalled by inactivation of Cd c Kinase, which leads to dephosphorylation of the proteins that were phosphorylated and the initiation of mitosis resulting in the exit from mitosis and the reformation of interphase nucleus.

The initial step in reformation of nuclear envelope is the binding of vesicles formed during nuclear membrane breakdown to the surface of chromosomes. This is followed by reassembly of the nuclear pore complexes, reformation of nuclear lamina and chromosome condensation to form a complete single nucleus. The nucleolus too reforms and transcription of rRNA genes begins, completing the return from mitosis to an interphase nucleus.

Cytokinesis

The completion of mitosis is usually accompanied by Cytokinesis, giving rise to two daughter cells. Cytokinesis in animal cells is mediated by a contractile ring of actin and myosin II filaments that form beneath the plasma membrane. The cell is cleaved in a plane that passes through the metaphase plate perpendicular to the spindle. Cleavage proceeds as construction of actin - miosin filaments pulls the plasma membrane inward, eventually pinching the cell in half.

The mechanism of cytokinesis is different in higher plants, which are surrounded by rigid cell walls. Rather than being pinched in half by a contractile ring these cells divide by forming new cell walls and plasma membranes inside the cell.

Thus, at the end of mitotic division, a cell gives rise to two daughter cells which contain the same number of chromosomes as that of the parent cell. These cells again grow in size and undergo mitotic division, the net result being the growth of the organism.

Meiosis

The term Meiosis was coined by J.B.Farmer in 1905. Meiosis is a reductional division that results in the formation of four daughter cells. Each daughter cell contains half the number of chromosomes as that of the parent cell. These daughter cells are thus haploid and also called a reduction division. Mitosis generally takes place in somatic cells but meiosis always occurs in cells of the reproductive organs resulting in the formation of haploid gametes. These gametes again fuse and form diploid zygotes (Fig. 5.4).

Meiosis can be classified into 3 types:

(a) Terminal
(b) Intermediate
(c) Zygotic meiosis.

(a) *Terminal meiosis:* is also called gametic meiosis and is found in animals and few lower plants. In this type of meiosis the meiotic division occurs just before the formation of gametes.

(b) *Intermediary meiosis:* In Intermediary meiosis the reductional division takes place at some intermediate time between fertilisation and the formation of gametes. This is also called sporic meiosis. This is characteristic for flowering plants.

(c) *Initial or zygotic meiosis:* occurs in some algae, fungi and some diatoms. Here the meiotic division occurs immediately after fertilisation.

Meiosis is a specialised kind of cell cycle that reduces the chromosome number by half. The exact replication and splitting of each chromosome in two identical parts and their subsequent separation into two cells would not ordinarily lead to any change in chromosome number between parent and daughter cells. In organisms where cells are always formed by sexual means, i.e. mitosis, the number of chromosomes would remain constant between generations. In sexually reproducing organisms where a zygote is formed by fertilisation of male and female gametes (haploid), the embryonic cells resulting from the fusion would have double the number of chromosomes than the parent cell, if no reduction in number of chromosomes occurred in the sex cell formation. But it is very clear that chromosome number always remains constant from generation to generation and so no such doubling occurs at any time. In the process of evolution, cells, i.e. gametic cells have somehow adapted to a mechanism of regularly reducing chromosome to half during sex cell formation. Meiosis is a simple process by which the chromosomes are separated during the formation of sex cells and their numbers reduced from diploid to haploid condition. Fertilisation then marks the event in which two haploid nuclei join to reform a diploid cell.

The number, as well as the size and shape of the chromosomes of a species, is called its karyotype and is usually constant. In diploid cells, each individual chromosome usually has a pairing mate called the homologue and both of them are called homologous chromosomes, one coming from each parent. In the absence of such homologous chromosomes the reduction division would result in random separation of chromosomes. This would result in haploid daughter cells but still lacking at least one essential chromosome. The reduction in chromosome number is accomplished by two sequential rounds of nuclear and cell division called Meiosis I and Meiosis II. which follow a single round of DNA replication. Each of these divisions is further divided into various stages as follows:

Meiosis I

Meiosis I, like mitosis, maintains after 'S' phase has been completed and the parental chromosomes have replicated to produce identical sister chromatids. The pattern of chromosome segregation, in meiosis I, is however, dramatically different from mitosis. During meiosis I, homologous chromosomes first pair with one another and then segregate to different daughter cells. Sister chromatids remain together, so completion of meiosis I results in the formation of daughter cells containing a single member of each chromosome pair consisting of two sister chromatids. The pairing of homologous chromosomes after DNA replication is only a key event underlying meiotic chromosome segregation, but also allows recombination between chromosomes of paternal and maternal origin. This critical pairing of homologous chromosomes takes place during an extended prophase of meiosis I which is divided into 5 stages. Leptotene, Zygotene, Pachytene, Diplotene and Diakinesis, on the basis of chromosome morphology.

Leptotene

The leptotene is first of the meiotic stages that differs from the previous interphase. The chromosomes first appear as long, slender threads, with many bead like structures called chromosomes, along the length. In some plants the chromosomes are clumped to one side of the nucleus. Such a condition is called synizesis. In some animals (many insects) they appear polarised with their ends drawn together toward that portion of the nuclear membrane close to the centriole. The initial association of homologous chromosomes is thought to be mediated by bass pairing between complementary DNA strands during this stage.

Zygotene

During this stage homologous chromosomes appear to attract each other and enter into a very close association in a zipper like fashion. This process is called synapsis. During this stage a zipper like protein structure called the synaptonemal complex is formed along the length of the paired chromosomes. This complex keeps the homologous chromosomes closely associated and aligned with one another. The pairing between homologous chromosomes is highly specific and occurs between all homologous chromosomes.

Pachytene

The Pachytene is a stage of progressive shortening and coiling of the chromosomes that occurs once zygotene has been completed. At this stage the two sister chromatids of a homologous chromosome are associated with the two sister chromatids of their homologous partner. The group of 4 chromatids is known as a bivalent or a tetrad. In this stage a series of exchanges of genetic material can occur or has already occurred between two non-sister homologous chromatids. The chromosomes may be linked at the sites of crossing over, giving the appearance of a typical chiasmata. The synaptonemal complex is also clearly seen.

Diplotene

At this stage, the synaptonemal complex disappears and each chromosome now acts as though it were repulsing its closely paired homologue, especially near the centromere. Distinctly visible separations occur between homologous chromosomes except for specific regions where an actual physical crossing over appears to have taken place between homologous chromatids. These are the chiasmas, which appear to be the only remaining force holding each bivalent together until metaphase.

Diakinesis

At diakinesis, coiling and contraction of the chromosomes continue until they are thick, heavy staining bodies. In this process the bivalents usually migrate close to the nuclear membrane and become evenly distributed. The nucleus disappears. The nuclear membrane slowly disintegrates and the bivalents rapidly attach themselves by their centrosomes to the rapidly forming spindles.

Metaphase I

In this phase the chromosomes reach their most condensed state and appear relatively smooth in outline. The chiasmata that had appeared during the diplotene now move towards the end of each chromosome. This process is called Terminalization. Terminalization leaves only the single terminal attach-ment between the formerly paired arms of homologous chromosomes. The kinetochores of sister chromatids are adjacent to each other and oriented in the same direction. The kinetochores of homologous chromosomes are pointed towards opposite spindle poles. Consequently, microtubules

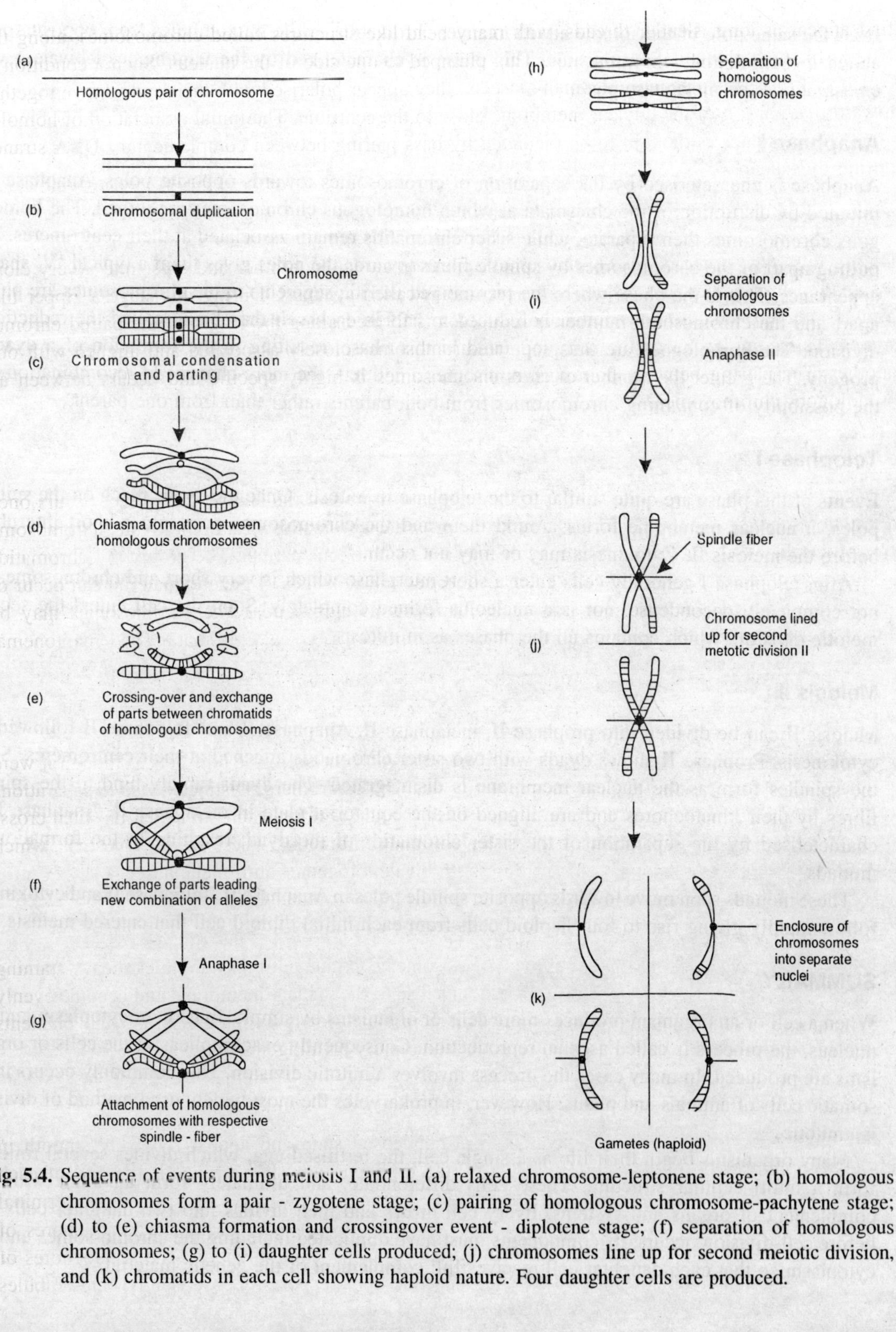

Fig. 5.4. Sequence of events during meiosis I and II. (a) relaxed chromosome-leptonene stage; (b) homologous chromosomes form a pair - zygotene stage; (c) pairing of homologous chromosome-pachytene stage; (d) to (e) chiasma formation and crossingover event - diplotene stage; (f) separation of homologous chromosomes; (g) to (i) daughter cells produced; (j) chromosomes line up for second meiotic division, and (k) chromatids in each cell showing haploid nature. Four daughter cells are produced.

from the same pole of the spindle attach to sister chromatids, while microtubules from opposite poles attach to homologous chromosomes. This phase is characterized by the alignment of bivalent with chiasmata on the metaphase plate.

Anaphase I

Anaphase is characterised by the separation of chromosomes towards opposite poles. Anaphase I is initiated by disruption of the chiasmata at which homologous chromosomes are joined. The homologous chromosomes then separate, while sister chromatids remain associated at their centromeres. The pulling apart of the chromosomes by spindle fibres towards the poles gives them a typical 'V' shaped appearance. This is the phase where the recombined genetic segments in the chromosomes are pulled apart and the chromosome number is reduced to half in each cell thus being called the reductional division. The homologues are thus separated in this phase, resulting in the formation of a dyad in progeny. The greater the number of chromosomes a cell has, the more chances of recombination and the possibility of containing chromosomes from both parents rather than from one parent.

Telophase I

Events of this phase are quite similar to the telophase in mitosis. Once the dyads reach on the spindle poles, a nuclear membrane forms around them and the chromosomes pass into a short interphase before the meiosis II. Cytokinesis may or may not occur.

After telophase I generally cells enter a short interphase which is very short and chromosomes do not completely decondense, nor is a nucleolus formed completely. Soon the cell enters the second meiotic division, which contains all the phases as in mitosis.

Meiosis II

Meiosis II can be divided into prophase II, metaphase II, Anaphase II and Telophase II followed by cytokinesis. Prophase II shows dyads with two sister chromatids attached at their centromeres. Soon the spindles form as the nuclear membrane is disintegrated. The dyads rapidly bind to the spindle fibres by their kinetochores and are aligned on the equatorial plate in metaphase II. Anaphase II is characterised by the separation of the sister chromatids of the dyad, resulting in the formation of monads.

These monads soon move towards opposite spindle poles in Anaphase II, Telophase II and cytokinesis follow rapidly giving rise to four haploid cells from each initial diploid cell that entered meiosis.

SUMMARY

When a cell or an organism produces more cells or organisms by simple division of cytoplasm and the nucleus, the process is called asexual reproduction. Consequently exact replicas of the cells or organisms are produced. In many cases the process involves a mitotic division. This commonly occurs in all somatic cells of animals and plants. However, in prokaryotes the most widely used method of division is amitotic.

Many organisms begin their life as a single cell, the fertilised egg, which divides several times to form a multi cellular structure. These cells differentiate, and ultimately form an adult with full complement of organs and systems. Every cell grows and then divides into two daughter cells and before cell-division occurs, its components must have duplicated including the chromo-somes and the cytoplasm so that each daughter cell receives full complement of the genetic material.

In most eukaryotic cells the actual process of cell division lasts for less than 1-hour, but the preparation for it may require a longer time during interphase. The overall events of the cell cycle are under the control of multiple gene products, which regulate the coordination of various events.

EXERCISE:

1. What is the significance of interphase?
2. Explain in detail various stages of mitosis cell division.
3. What is the significance of meiosis?
4. Explain in detail various stages of meiosis.
5. Write in detail what do you know about cell cycle?
6. Write short notes on:
   (a) pachytene (b) leptotene
   (c) xygotene (d) diplotene

6

Cell Motility

Cell motility may involve movement of the entire cell or a portion of it, to the benefit of the organism. This is a biological phenomenon displayed to enable many roles such as feeding, digestion, reproduction, circulation and protection. Flagella, cilia and particularly a muscular system are the means of locomotion for a cell or a unicellular organism.

Biological movements are also performed by cytoplasm of the cells such as cytoplasmic streaming in plant cells or cyclosis in Amoeba.

AMOEBOID LOCOMOTION

As there is no exterior cell wall in animal cells, finger like blunt protrusions from the cell may project, which help in locomotion and also feeding.

In Amoeba, which is a unicellular Protozoan, such protrusions are called pseudopodia which are formed at any point on the surface. Pseudopodia are temporary processes formed when an Amoeba is moving on a solid substratum.

It is suggested that actin filaments and myosin molecules are found in Amoeba. When they interact, actin-myosin complex is formed, providing the basis for locomotion. Due to this interaction, contraction is possible with continuous hydrolysis of ATP.

Various hypotheses have been postulated for explaining the formulation of pseudopodia and locomotion in Amoeba. Amoeba has a layer of ectoplasm that is more rigid and gel-like, whereas endoplasm is diluted and sol-like. When ectoplasm contracts in a certain region, the liquid endoplasm streams towards the region, which can be observed under a polarised microscope. When the endoplasm is propelled in the ectoplasmic tube, it also changes its viscosity and becomes rigid. Automatically contraction is created in the direction opposite to that of streaming of endoplasm.

According to Mast, during the formation of pseudopodium, the plasma membrane gets attached to the substratum followed by a local and partial liquefaction in the plasmagel at that point. Rest of the plasmagel flows out in that area to produce a bulge (Fig. 6.1). At the posterior end, contracting plasmagel is converted into plasmasol and anteriorly, an ectoplasmic tube is continuously regenerated by gelation of plasmasol.

The sol-gel interconversions are actually the contraction-relaxation events, which are enforced by osmotic pressure and other ionic changes. It is considered that ca^{++} ions play an important role in regulating the amoeboid movements. Free ca^{++} ions in the plasmasol induce contraction and bring about conversion of sol into gel.

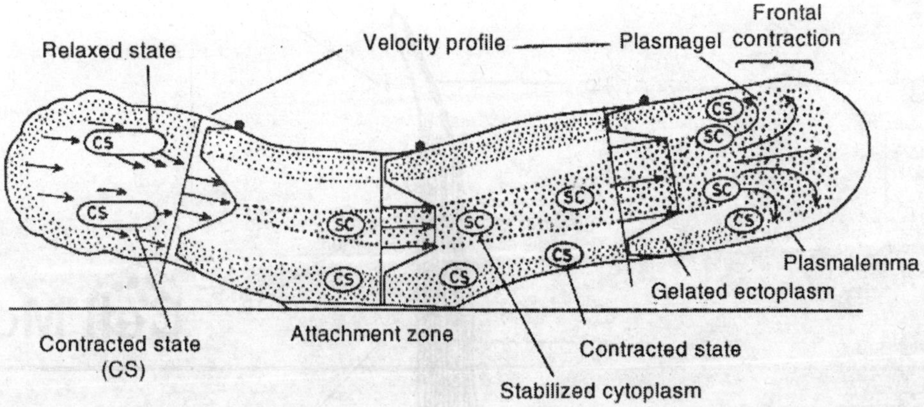

Fig. 6.1. A proposed scheme to show amoeboid locomotion. The cytoplasm on reaching the frontal end undergoes contraction and merges with the plasmagel which moves posteriad as ectoplasmic gel. The gel is then converted to sol which is then converted into relaxed state

Flagellar and Ciliary locomotion

With the help of specialised locomotor organelles on the surface, viz., flagella and cilia, many unicellular organisms exhibit locomotion.

Flagella are long, slender, thread-like projections from the surface of the cell. The base of the flagellum is anchored in the cytoplasm on a motion-controlling kinetosome region or granule. A cell usually has one flagellum but in many cases there may be more anchored in the kinetosome granules.

Cilia are shorter and each cilium has its own kinetosome at its base (Fig. 6.2). A cilium is about 1 to 15μ long, whereas a flagellum can be about 150μ long.

Cilia and Flagella are made up of a central core of cytoplasm surrounded by a double membrane which is actually the extension of the cytoplasm. Inside the cytoplasm pairs of elongated fibres are arranged along the periphery and two unpaired elongated fibres are present in the centre. In all there are 9 paired and 2 unpaired fibrils attached to a basal plate.

The cilium originates from a basal body or kinetosome which lies embedded in the ectoplasm. The tip of the flagellum or cilium tapers to a point where the number of fibrils are reduced. In certain cases the basal region contains a single secondary fibre. In some cells fine ciliary rootlets are also found arising from the basal granule. Cross section of cilium shows that the axial microtubular structure of axonema consists of nine pairs of longitudinal tubules, arranged usually around two central unpaired filaments. The central tubules may or may not always be present, hence it is suggested that the essential motile elements are the peripheral tubules. The paired peripheral microtubules are ellipsoidal, whereas the central ones are circular.

Each paired microtubule or doublet has a subfibre A and subfibre B, out of which subfibre B is larger. Each subfibre A gives rise to dynein arms, all oriented in a clockwise direction.

Subfibre A of the doublet gives rise to radial links or spoke like structures and they reach the central sheath of unpaired fibres (Fig. 6.3). The radial links end up in a dense knob like structure. The dynein arms contain dynein or a high molecular weight ATPase, requiring Mg^{2+} and Ca^{2+} for its activity. The basic mechanism of ciliary or flagellar motion, as proposed by Gibbons in 1977, is due to interaction between tubulin and dynein.

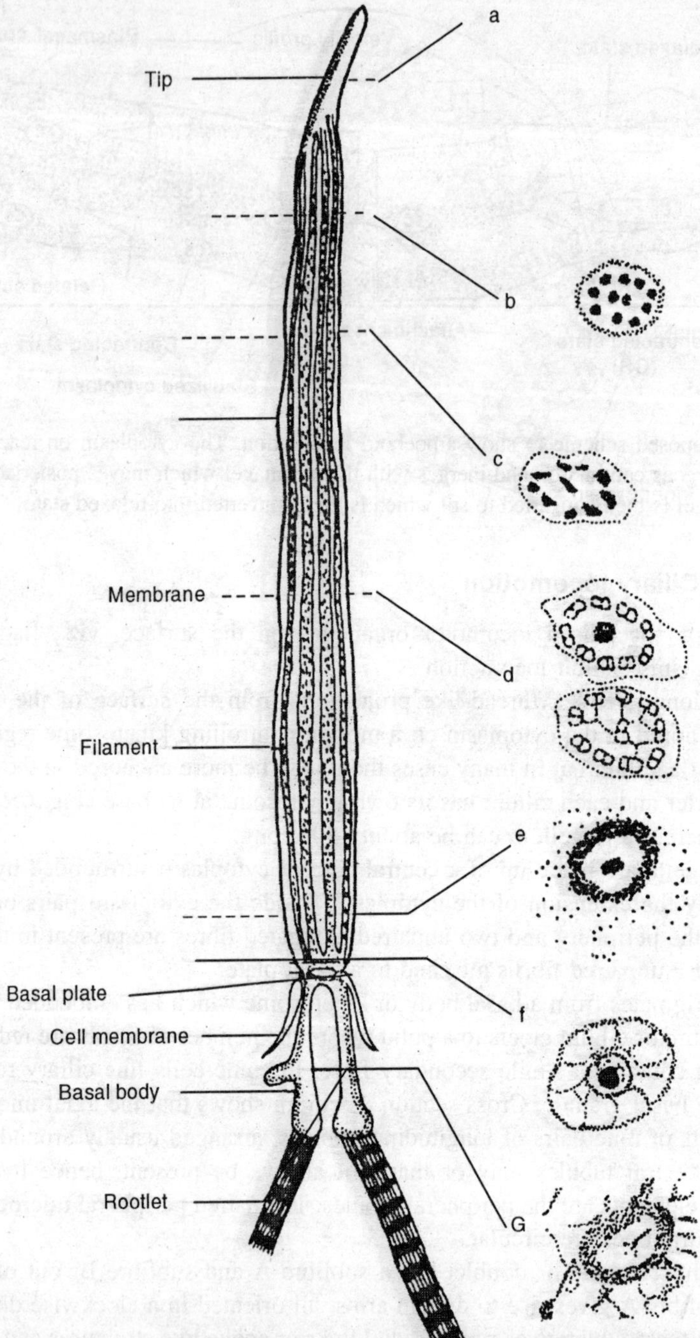

Fig. 6.2. The cilium and ciliary apparatus. Internal structure of cilium shown at different levels. Basal plate from which the cilium arises lies embedded in the ectoplasm. The cilium consists of 9 pairs and 2 unpaired tubules attached to the basal plate

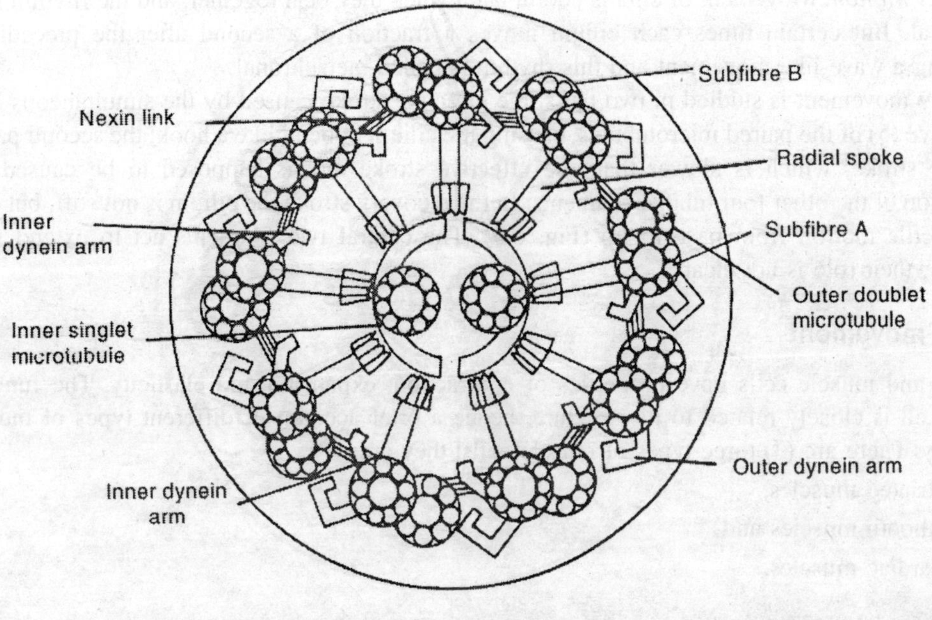

Fig. 6.3. Cross section of the cilium showing subfibres A and B, and dynein arms

Basal body or kinetosome is similar to the centrioles of the mitotic spindle. The basal body in a cross section is seen to consist of 9 groups of tubules arranged in a circle. Each group of microtubule is a triplet designated A, B and C tubules.

The tubules are connected with each other and also to the central fibre through radial connections (Fig. 6.4).

All tubular fibres are enveloped in a membrane, and from the basal body in many cases two thin filaments or rootlets are found to extend. From the basal plate of the cilium two dense perpendicular processes arise which seem to originate from the triplets. These pro-cesses have been termed the basal feet, which are composed of microfilaments.

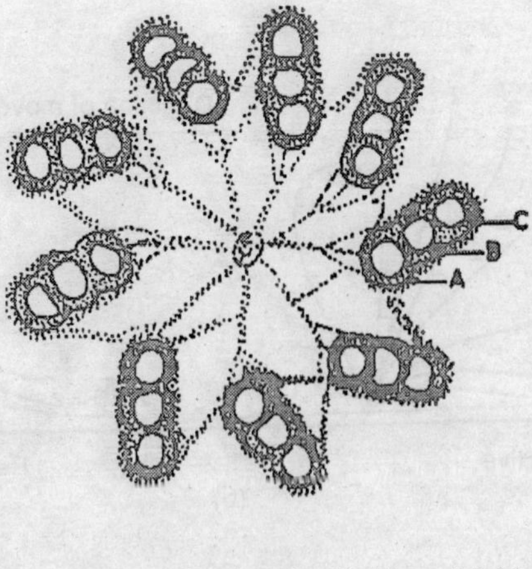

Fig. 6.4. Cross section of the basal body of cilium (highly magnified)

Ciliary Motion: Movement of cilia is coordinated when they beat together, and the rhythm is called isochronal. But certain times each cilium moves a fraction of a second after the preceding one, producing a wave-like movement and this rhythm is called metachronal.

Ciliary movement is studied in two parts: the effective stroke caused by the simultaneous contraction of five (5) of the paired microtubules, causing the cilium to bend like a hook; the second part is the recovery stroke, which is slower than the effective stroke and is supposed to be caused by the contraction of the other four tubular filaments. In the recovery stroke the cilium is not stiff, but exhibits a contractile motion from base to tip (Fig. 6.5). The central two filaments act to extend support. However, their role is not clear.

Muscle movement

Muscles and muscle cells have properties of contraction, expansion and elasticity. The function of muscle cell is closely related to its structure, hence a brief account of different types of muscles is necessary. There are (3) three types of muscle cells; they are:

(i) Striated muscles,

(ii) Smooth muscles and,

(iii) Cardiac muscles.

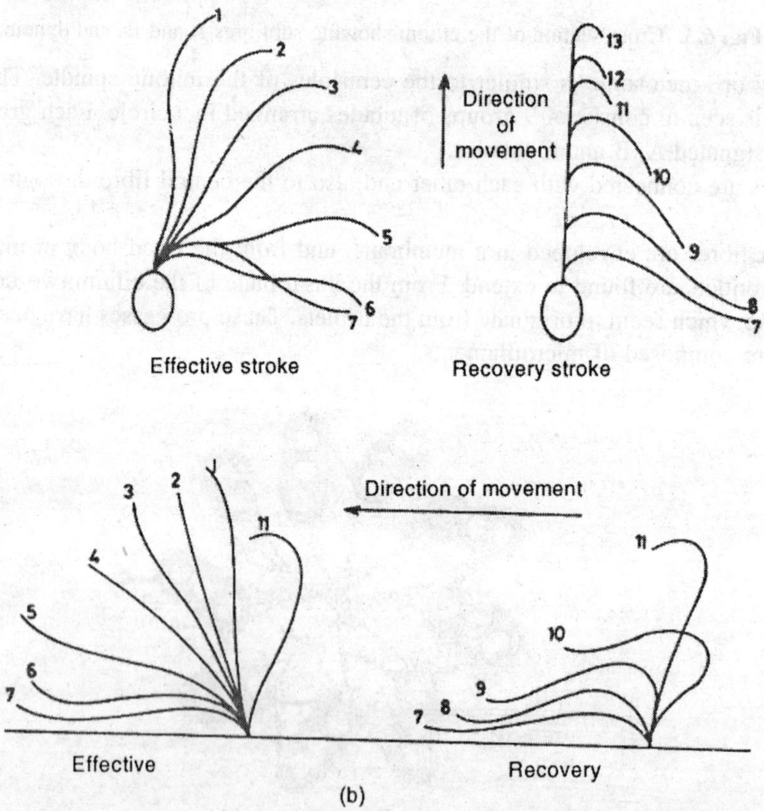

Fig. 6.5. Flagellar and ciliary movement: (a) Flagellar movements during effective and recovery stroke; (b) successive ciliary motions produced by bending movements. Bending motion is produced by making and breaking of cross bridges between adjacent microtobular doublets

Structure and function of striated muscle cells: These cells are composed of multinucleate muscle fibres. Each fibre consists of a semi-fluid sarcoplasm, containing many longitudinal myofibrils, called fasciculus. The membrane enclosing each fibre is called sarcolemma, beneath which are located numerous nuclei is scattered. A live myofibril transparent, shows cross-striations with alternating light and dark bands, under polarised microscope.

Myofibrils are basic contractile units which in a relaxed state show dark and light bands due to the concentration of myofilaments. (Fig. 6.6).

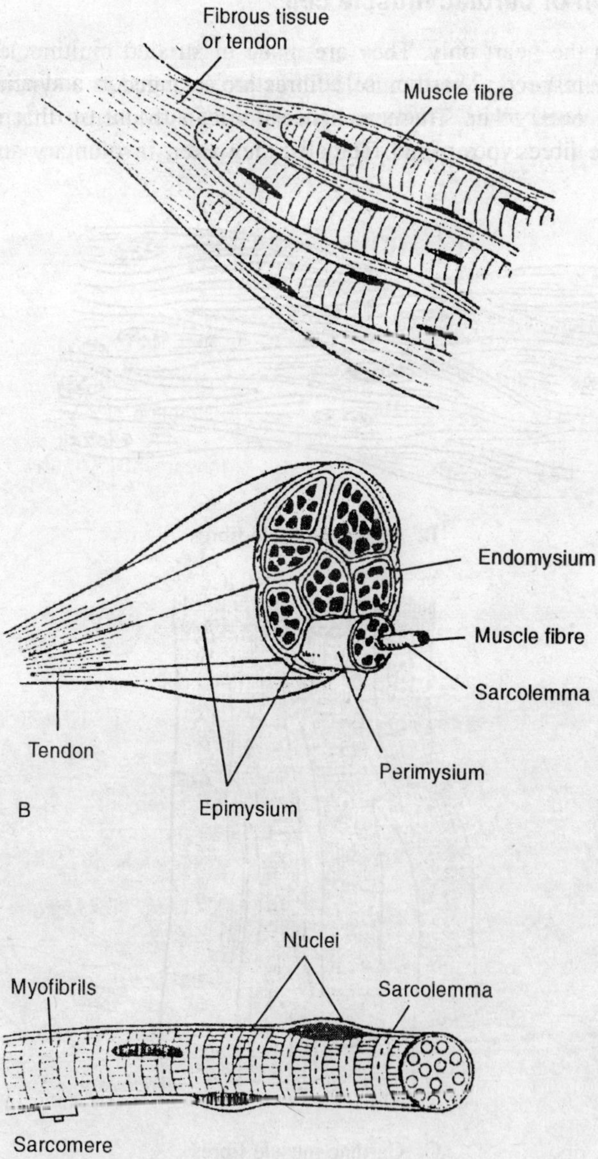

Structure and function of smooth muscles cell

They are also called unstriated or involuntary and are devoid of any cross-striations. They are spindle shaped cells with long-tapering ends and a centrally placed large nucleus. They are generally found in the wall of internal organs such as digestive tract, respiratory passage, urinary bladder, arteries and veins. They are slow in their contractile behaviour and do not come under voluntary control. Although these cells contain actin and myosin proteins, their mechanism of contraction is still not completely understood (Fig. 6.6).

Structure and function of cardiac muscle cell

These cells are found in the heart only. They are made of striated multinucleate fibres, resembling skeletal muscles in many respects. These muscle fibres are arranged in a syncitial fashion and do not appear to be fused with each other. These are specialised in the sense that stimulation of cardiac muscle causes all muscle fibres to respond. They are, however, involuntary and innervated by autonomic nerves (Fig. 6.6).

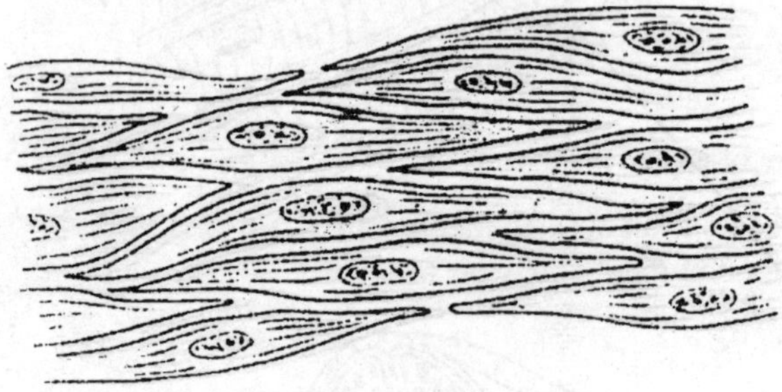

B. Smooth muscle fibres

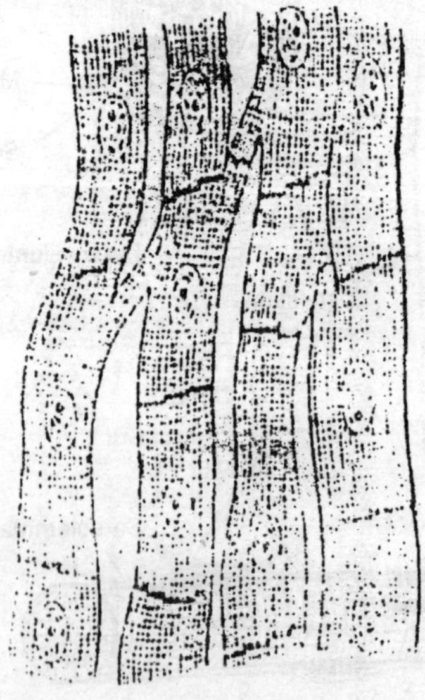

C. Cardiac muscle fibres

Fig. 6.6. Striated muscle fibres and their connective tissue Longitudinal section with some connective tissue approved. A. Cross section B. Single fibre C. Sarcomere

Nerve cell structure and functions

A nerve cell or the "*neuron*" is the functional unit of conduction and transmission of nerve impulses.

The neuron is a microscopic structure consisting of the main cell body, dendrites in the form of out growths and a long axon terminating into axon fibres ending in a bulb-like structure (Fig. 6.7). The axon is a hollow cylinder filled with axoplasm, which differs in chemical composition from the surrounding fluid. It is along this axon an impulse travels. The main cell body with its nucleus is needed to generate an impulse. The axon can be distinguished from the dendrites as it is longer and carries neurosecretory vesicles. Neuron is covered by a myelin sheath which acts as an insulator (Fig. 6.7).

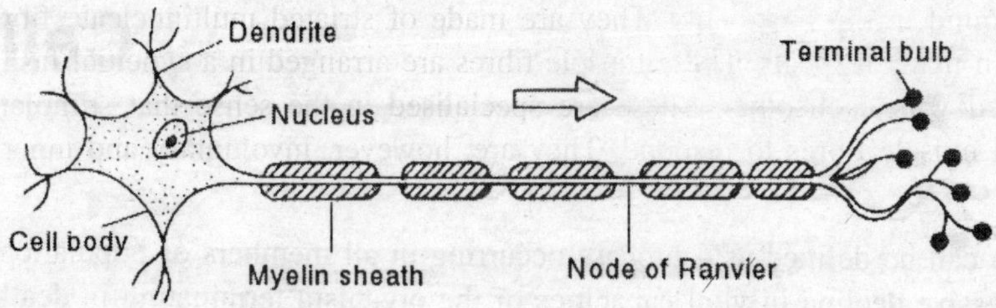

Fig. 6.7. Structure of a neuron show myelinated axon terminating in axon fibres

The neuron when excited develops an action potential, which is a result of profound changes in the electrical property of the nerve membrane. The events that follow excitation are conduction of impulse and the response evoked by the stimulus.

SUMMARY

Cell locomotion may involve movement of the entire cell or a portion of it, to the advantage of the organisms. It is a phenomenon that is displayed to accomplish many roles such as feeding, protection, digestion, reproduction, and circulation. Flagella, cilia and particularly a muscular system are the means of motion for a cell or an organism, and this motor ability is used mainly to capture food in two major ways. Biological motions are also performed by cytoplasm of the cells such as cytoplasmic streaming in plant cells or cyclosis in Amoeba. The cytoplasmic motion is a great aid in the transport of materials. Once the capacity of motion is acquired, it may serve not only in feeding but also in protecting them from potential dangers.

EXERCISE:

1. Discuss various types of locomotary organelles, and their utility to organism.
2. Describe the ciliary apparatus and the mechanism of ciliary motion.
3. Explain in detail amoeboid locomotion with supporting hypotheses.
4. Explain in detail about different muscle cells their structure and function.
5. Write Shortnotes on:
   (a) striated muscle.
   (b) smooth muscle.
   (c) cardiac muscle.
   (d) nerve cell.

7

Cell Senescence and Programmed Cell Death

Cell Senescence can be defined as a process occurring in all members of population, after maturity, involving progressive decline in vital capacities of the organism terminating in death. Senescence is also considered as death.

The science that studies biological causes of senescence is called gerontology.

Senescence (Ageing) follows cessation of growth. At the cellular level senescence can be studied on the basis of three processes:

(a) Decline in the final efficiency of non-dividing highly specialised cells such as neurons and muscle cells to the possible extent.

(b) Progressive stiffening with age of the structural proteins – such as collagens.

(c) Limitation imposed on cell division as revealed by the studies on fibroblast producing collagen and fibrion.

Mechanisms of Cell Ageing: It can be suggested that normal cells have a finite capacity for replication, and this finite limit is rarely reached in vitro but is, of course, demonstrable in vitro. The functional losses that occur in cells prior to their loss of division capacity produce age changes.

According to L. Orgel (1963), cellular ageing results from impaired specificity of the translation step in protein synthesis. This hypothesis has been confirmed to a great extent.

THEORIES OF SENESCENCE

Many theories related to cell senescence depending upon the different types of nucleic acid and nucleoprotein structures are described.

A. *Change in Nucleic Acids Quantity*: Loss of DNA or RNA per cell per organ could explain decreasing functional efficiency with increasing age. Histological findings point to progressive loss with age of certain irreparable types of cells.

B. *Mutation theory*: The effect of mutations is to bring about the synthesis of faulty messenger RNA and thus, in turn, faulty proteins which are unable to fulfil their biological functions or can do so only imperfectly. Due to faulty proteins, impaired specificity of the enzymes are formed in the translation mechanism, according to Orgel.

There are many ways in which protein synthesis - information content can be changed, falsified or diminished. Results of many biologists' observations are mentioned below:

(1) In ageing cell replacement of defective molecules of metabolic DNA becomes impossible, therefore defective molecules accumulate; as a result of many faulty DNA molecules in the cell, functional impairment takes place.
(2) Ageing is attributed to loss of repetitive information.
(3) The number of methyl group (5′ methyl cytosine) in DNA decreases with age, thus DNA information content gets modified and affects protein synthesis.

C. Modifications in protection regulatory mechanisms: According to Orgel, ageing of cells may ensue as a result of cumulation of trans-cription and translation errors in protein synthesis. Errors leading to reduced specificity of an information handling enzyme lead to an increasing error frequency. Such processes are cumulative. It is a question whether cellular ageing may be due to failure or switchover regulatory processes. Could there be an ageing programme? As an answer to this DNA pool in old age has been observed, and a hypothesis was formulated that each species has species specific programme that serves to maintain and prolong useful life. It is also proven that there is a shift from, arginine-rich to a lysive rich histone composition in the liver of old animals.

FREE RADICAL THEORY OF AGEING

Ageing process may be divided into two categories of cumulative degradative changes.

(i) wide damage produced by a variety of means, such as autoimmune reactions, ionizing radiations and smog.
(ii) alterations in the so called biological clocks-determine the maximum life span of an individual cell.

Free radical reactions lead to a variety of products

Initiation	RH+O	-cu-	R+H O
Propagation	R+O	-	RO
	RO+RH	-	R+ROOH
Termination	R+R	-	R:R

The high reactivity of free radicals is due to the presence of a free electron. Due to the magnetic moment, associated with a free electron, free radicals can be detected at low concentration using the technique of electron spin resonance (ESR) spectroscopy.

Free radical reactions rate, involving molecular oxygen is accelerated by catalysts such as copper, iron and manganese; and inhibited by antioxidants such as vitamin E, butylated hydroxytoluene, and 2-mercapto ethylamine (2-MEA) which are capable of removing intermediate free radicals. These compounds are expected to minimize deleterious effect of free radicals. E-vitamin is a natural antioxidant which has a modest beneficial effect on the life span as it decreases the rate of free radical formation.

SENESCENCE AND IMMUNOLOGICAL SURVEILLANCE

Senescence process is basically related to the fate of the less differentiated mesenchymal cells responsible for the functions of defence and repair with lymphocytes and fibroblasts. Like other somatic cells, these cells are prone to mutations. Though genetic mutations in the somatic cell may modify any cellular activity, they cause two major changes.

(a) Against the changed cells, altered antigens generate immune responses and
(b) Mutation within lymphycytes induces a change in torlerance to self components.

SOMATIC MUTATION HYPOTHESIS

Every living species has certain life span, beyond which genetically controlled programmes become less effective until death. Most vulnerable material to age related damage is the genetic material, i.e. DNA. Changes in the DNA base sequence alter the template and, as a result, affect the regulatory and metabolic capacities of the cell. Cell genome damage may occur in several ways, such as radiations or multiplying errors caused by unmatched nucleotides. Somatic mutation continuation involves error in the structure of enzymes connected with protein synthesis, causing an irreparable damage that will lead to failure of cell mitosis.

THYMUS FUNCTION

Thymus is a very important "Biological Clock" which allows phenotypic expression of genetically decided age. It is a known fact that serum antibodies are produced only by B-lymphocytes and most natural antigens can stimulate antibody production exclusively by the cooperation of T and B immunocytes. T-immunocytes (thymus determined lymphocytes) without B-cell co-operation will lead to delayed hyper sensitivity, for homograft rejection and for medical examination important segment of immunity against viral and mycobacterial infections. Immunological inadequacies of old age involves weakness of T-cell rather than B-cell function. So, due to losing flesh of thymus in middle age, T-cell loses to incorporate effective immunocyte clones against antigens which were previously not encountered.

IMMUNE SURVEILLANCE

Somatic mutations are neither continuous nor heritable changes in a cell line. The development of new plastic change, benign or malignant, must be considered as somatic mutation. Such modifications are sometimes involved with some degree of antigenicity.

Common types of cancer can be considered as old age disease in which any directly influencing carcinogenic agents are very common to have influence on incidence. Senescence may be presumed to act in two ways:

(1) By allowing time for any mutation to occur.
(2) By a progressive weakening of immune surveillance with age.

SENESCENCE OF CONNECTIVE TISSUE

Gross age related changes are observed in connective tissues of the body. All the forms of connective tissues, viz., dermis, tendons, cornea, vascular walls, cartilages, bones, etc. are derived from embryonal mesenchyma. All these connective tissues possess large intercellular spaces filled with collagen, elastin, proteoglycans and structural glycoproteins, collectively considered as intercellular matrix macromolecules (IMM).

In the beginning of embyyonic life proteoglycons and structural glycoproteins participate actively in differentiation process and later this role is taken over by collagen and elastin. In adult life, collagen functions whenever necessary, whereas elastin is suppressed. Collagen synthesis reduces rapidly at an advanced stage.

SENESCENCE OF ELASTIN TISSUE

Elastin fibres play an important role in elasticity and normal tone of skin and blood vessels. Quantity of these elastic fibres deteriorates with age in skin, blood vessels, particularly during arteriosclerosis and ageing disease. Degradation of elastic fibres start usually at a relatively young age, but at the age of 45 years, the degradation accelerates.

AGEING AND TERMINATION OF SYNTHESIS PROGRAMME

Translation and transcription regulations at the genetic level determine the speed of ageing process, and if this control is upset, the cells may synthesise wrong molecules or stop synthesising at all. This process may be either due to exhaustion of genetic programme or to the collection or errors.

Further, another mechanism may also contribute to senescence of ageing; which is dependent upon the cellular micro and macro environment, on nutrition, on physical exercise and balanced diet etc. Avoidance of regular exercise damages healthy functioning of muscle tissues.

"APOPTOSIS"

Multicellular organisms, homeostasis is maintained through a balance between cell multiplication and cell senescence(Cell Death). Physiologically cell death occurs mainly through a process apoptosis.

Recent evidences suggests that alterations in cell survival contribute to pathogenesis of a variety of human diseases including cancer, viral infections, autoimmune diseases, neurodegenerative disorders and AIDS. Treatments established to alter the apoptotic threshold may cure some of these diseases (Thompson 1995).

Apoptosis is a kind of programmed cell death which plays an important role during development, homeostasis, and in several diseases like cancer, acquired immuno deficiency syndrome (AIDS) and neurodegenerative disorders. It operates through the activation of a cell intrinsic suicide program. Its important machinery appears to be present in most mammalian cells at all times, but the activation of the suicide program is regulated by many different signals that originate from both intracellular and extracellular milieu (stellar, 1995). Some components of apoptotic program are established among worms, insects, and vertebrates (Stellar 1995).

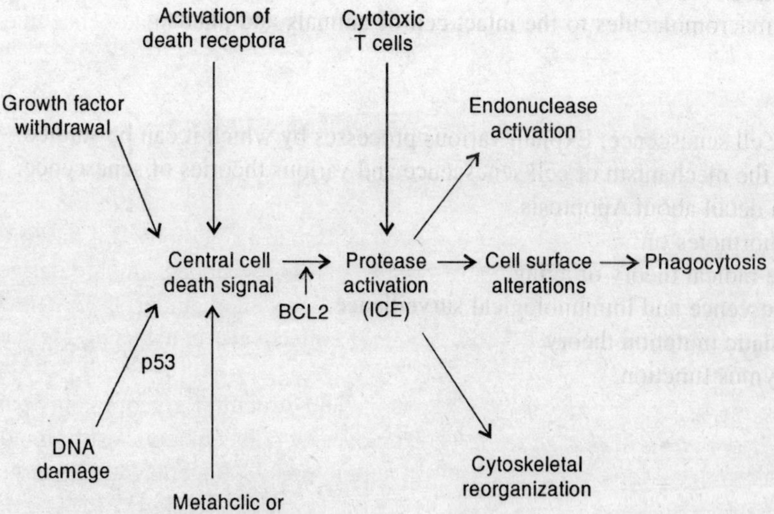

Fig. 7.1. Model for the regulation of apoptosis. The major end point of apoptotic cell death is the removal of the dying cell by phagocytosis (After Thompson, 1995).

Homeostasis in higher organisms is regulated not only by the multiplication and differentiation of cells but also by cell senescence. Cell senescence during embryo genesis, metamorphosis, endocrine dependent tissue atrophy, and normal tissue turnover is termed programmed cell senescence. Most programmed cell senescence occurs by apoptosis which includes condensation and segmentation of nuclei, cytoplasm and extensive fragmentation of chromosomal DNA into nucleosome units.

Most animal cells are capable of self destruction by this intrinsic cell suicide program. Execution of this apoptosis is often associated with morphological and biochemical changes. During apoptosis, nucleus and cytoplasm condense and the dying cell fragments into membrane bound apoptotic bodies, which are quickly phagocytosed and digested either by macrophagus or by neighbouting cells. Dead cells are thus quickly removed, leakage of their poisonous or dangerous contents is avoided. Apoptosis is different from necrosis which is a pathological form of cell death that results from cellular injury. Apoptosis is usually associated with the activation of nucleases that degrade the chromosomal DNA first into large (50-300 kilobases) and subsequently into very small oligonucleosomal fragments.

Apoptosis is very important for the development and homeostasis of metazoan animals. Besides the beneficial effects of cell death, the inappropriate activation of apoptosis can contribute to several diseases, including AIDS and ischemic stroke (Fig. 7.1).

Certain cysteine proteases are involved in causing apoptosis. The actual cause and mechanism of apoptotic death is still unknown. But there are several similarities between apoptosis and the cell cycle implying that apoptosis and mitosis may be mechanically related or even coupled. There is also an extreme view of this idea that apoptosis may be an aberrant type of mitosis.

SUMMARY

The most important inviolable law that governs biological events of living cell is senescence or ageing. The process of ageing is an inevitable biological event that can neither be slowed nor stopped. The science that studies biological causes of senescence is called gerontology. Ageing processes are progressive and not reversible under physiological conditions. In living systems ageing occurs at all levels—from macromolecules to the intact cell of animals and plants.

EXERCISE:

1. Define Cell senescence; Explain various processes by which it can be studied.
2. Explain the mechanism of cell senescence and various theories of senescence.
3. Write in detail about Apoptosis.
4. Write Shortnotes on:
   (a) Free radical theory of aging.
   (b) Senescence and immunological surveillance.
   (c) Somatic mutation theory.
   (d) Thymus function.

Part B: Genetics

8

Structure of DNA and RNA

NUCLEIC ACIDS

The Nucleic Acids are divided into two categories the RNA and DNA. In different organisms the genetic information is stored either with in the DNA or the RNA. Example, *Bacterial Viruses* (Bacteriophages) ϕ x 174, λ, T_2, T_4, have DNA. F_2, Ms_2, R_{17}, and Q_β have RNA as genetic Material.

Animal Viruses

Simiam Virus 40, Mouse Polyoma and Rabbit papilloma, Herpes simplex, adenovirus have DNA. Rous Sarcoma in fowl, poliomyelitis, influenza, Reo virus have RNA.

Plant Viruses

T.M.V. and Tomatobushy stunt have RNA

The Nucleic acids: consists of nucleotides, as their building blocks, where, they are made up of Nucleoside and the phosphate group, the Nucleosides are further made up of pentose sugar + Nitrogenous base.

Pentose in RNA is ribose while in DNA, it is a 2′ Deoxy ribose. But the pentose occur in the Nucleotide in their β-furanase form. The PO_4 gr. is esterified usually at 5′ position and in the cell pH has a -ve charge. The Nitrogenous bases are linked to the 1′ Carbon by n-glycosyl linkage.

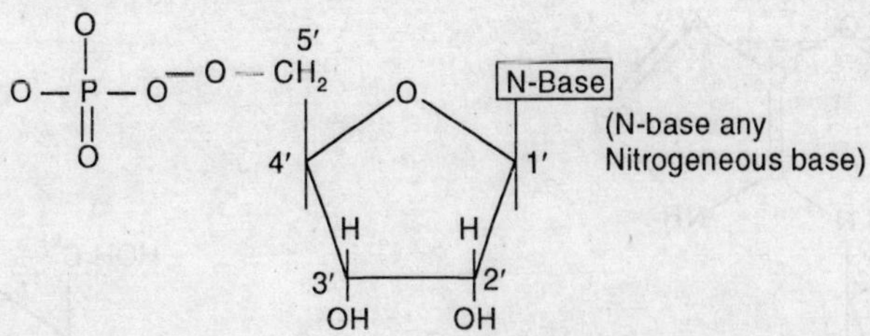

Nitrogenous bases are the derivatives of 2-parent heterocyclic compounds namely pyrimidine and purine. Derivatives of pyrimidine are cytosine and Uracil and thymine where the RNA consists of cyto sine and uracil while DNA has T and C. The purine derivatives are adenine and Guanine both of which occur in DNA and RNA.

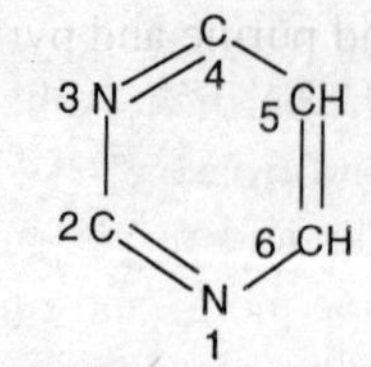

Pyramidine

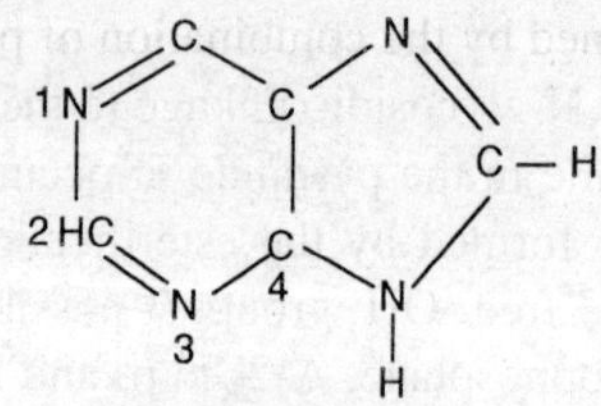

Purines

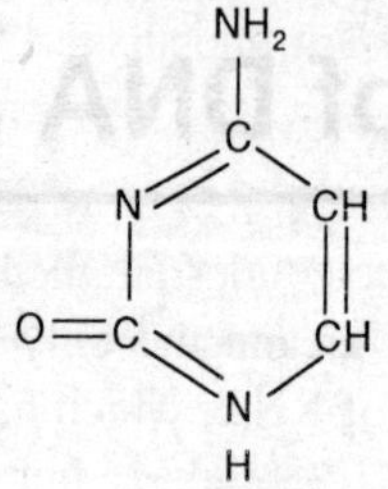

Cytosine

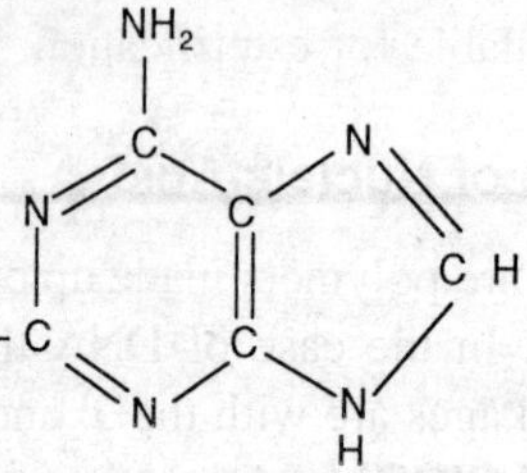

Adenine

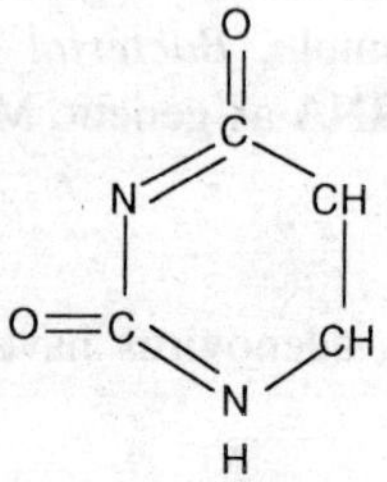

Uracil

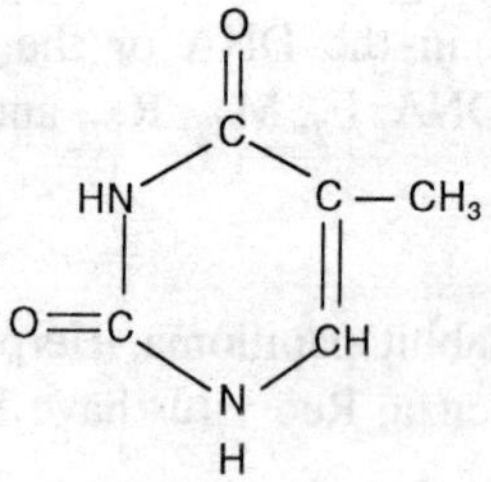

Thiamine

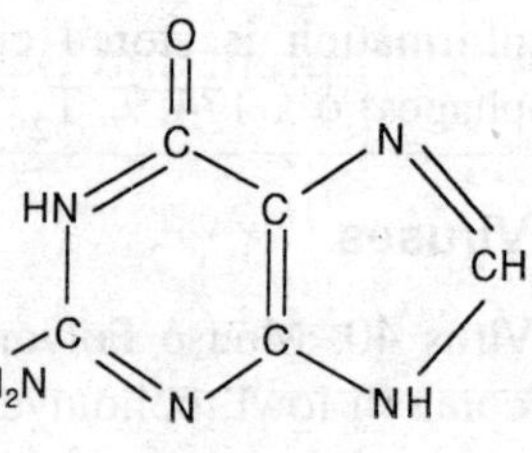

Guanine

only -CH_3 is the difference

Apart from these major bases some minor bases, which are methylated forms of the principle base and in some viral DNAs they may be methylated or glucosylated. Such bases have specific roles as signals. The pyrimidine bases usually undergo keto enol tautemeriun as shown below.

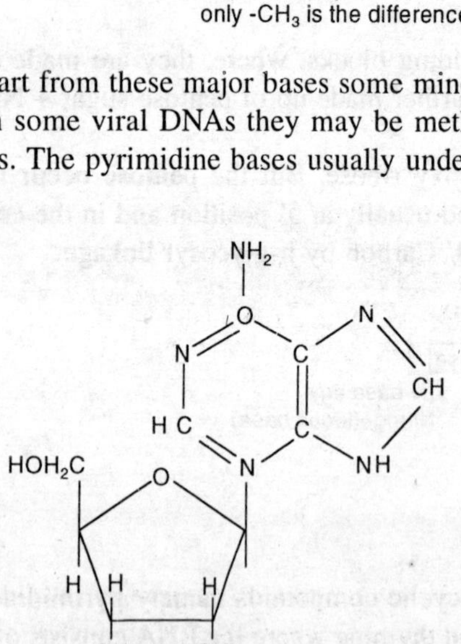

Adenosine

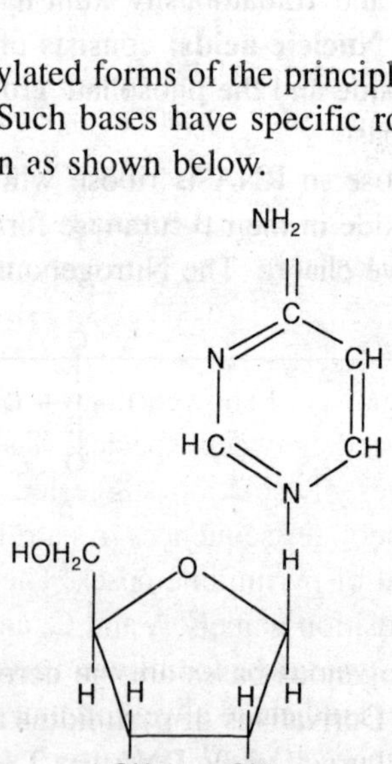

Cytidine

A nucleoside is formed by the combination of pentose sugar and purine and pyrimidine base, where the bases are linked by N-glycosidic linkage to the 1′ carbon. In the purine it is the 9th position, where the linkage occurs, while in the pyrimide, it occurs in 1st position of base.

The nucleotides are formed by the esterification of phosphoric acid on the nucleoside. Since the ribo nucleosides have 3 free 'OH' groups 3 possible ribonucleoside mono phosphates can be formed.

Adenosine 3′, Monophosphate, A-2′ m.p. and A-5′ m.p. The ribo nucleosides at the 5′ phosphate may be further phosphorylated to yield the 5′ di and tri phosphates. Eg: AMP, ADP and ATP.

However, no such formation, is seen with the deoxy ribonucleoside, which has only the 3′ and 5′ positions, available for esterification. They also form mono, Di Tri, phosphates.

1° Structure of Nucleic Acids

Nucleic acids are polymers made up of 100, ... 1000, ... many nucleotides coupled together by phospo-di-ester bond. In the case of DNA the 4 carbon has no 'OH' gr and the only availability for inter nucleotide linkages are with the 3′ and 5′. However in the case of RNA, which has the 'OH' gr, at the 2′ position, it is possible to postulate 2′, 5′ linkage in addition to 3′, 5′ linkage but result indicates, even in RNA it is the 3′, 5′ linkage.

Nucleic Acid Content and Base Composition

Organism	*Haploid DNA 10^{-12} g (Pico gram)*	*Purine (A.G.)* *A*	*G*	*Pyrimide* *T*	*C*	*A + G/ T + C*
Man	3.2	31	19.1	36.5	18.4	1
Cattle	3	28.1	22.2	21.2	22	1.03
Fruit fly	0.18	27.3	22.5	27.6	22.5	1
Corn	7.5	25.6	24.5	25.3	24.6	1
E. Coil	0.0047	24.6	25.5	24.3	25.6	1
Heroes(simplex)	0.00011	13.8	37.7	12.8	35.6	1.06
Lambda	0.000055	26	23.8	25.8	24.3	0.99
174 (Single standard)	0.00000284	24.7	24.1	32.7	18.5	0.95
TMV	0.0000033	29.3	25.8	26.8	18.1	1.23

Chorgaff et al showed that all the nucleotide bases were not present in equal amounts and that the ratio varied between species. The experiments conducted by him in conjunction with improved techniques of isolation suggested that the DNA were not a single molecule, made up of repeated tetranucleotides sequences. As a rule chorgaff had shown, that the purine bases (A + G) equalled to the sum total of pyrimidine bases. There was an equivalence between bases carrying amino gr. at the 6th or 4th position namely A and C, and the bases carrying keto groups namely T and G, the ratio of A to T, G to C was close to unity in various eukaryotic species. X-ray studies on DNA by Wilk-ins, Franklin and others indicated a well organized multiple stranded fibre about 22 A° in diameter, characterised by groups spaced 3.4 A° and a repeating unit every 34 A°. Taking into account the points, Watson and Krick proposed a double helical structure for DNA. According to them, the DNA molecule is 2-stranded and coiled like a rope, so that by permitting the ends to revolve freely, the two complementary strands can be separated. The coiling is helical, like a circular stair case that maintains the same diameter. The backbone is composed of phosphate + sugar linkage, the half-step of a strand extends to meet the half step of the complementary strand. There are single purine or pyrimidine bases, each step, therefore consists of a pair of base, termed base pair or complementary base pair. The steps are 3.4 A° a part, and each is turned 36° from the preceding one, so that a complete turn of 360° involves 10 stairs, or 10 base pairs, which is 34 A° long.

The rotation of this double helix around it's axis generates one major (wide) and one minor (narrow) grove. In each complete turn the minor groove may bind to some Histone proteins while the major groove binds to non histone proteins. The base pairs are held by 'H' bonding as shown in the figure.

Different Conformational forms of DNA

Form	Forms of DNA	Pitch (nm)	Residues per turn	Inclination of base Pair from horizontal
A	Na-salt 75% RH.	2.8	11	20°
B	Na-salt 92% RH.	3.4	10	0°
C	Lisalt 66% RH.	3.1	9.3	6°

Through the basic model put forward by watson and crick is close to the accepted structure of the DNA molecule in solution. The refined X-ray diffraction studies of wilking etal indicates three possible structures of DNA fibres. The B structure comes close to the original watson – crick model.

The A and C conformations are also right handed but differ in the pitch and in the number of bases per turn. In both A and C forms the bases are tilted. Their biological significance is not clear but the A conformation is believed to be close to the structure adopted by double stranded RNA and by DNA-RNA hybrid. Because of the presence of the extra 2-OH group, RNA appears to be unable to adopt the B confirmation. Thus during the formation of the template for making RNA, the DNA molecule must adopt the conformation.

Various organic solvents also force the DNA to change its conformation. Adjacent bases are related to each other by roll, slide and twist and the A, B and C forms are simply structure which exist across a continuation of values. The bases may slide relative to each other with the values of slide (σ) ranging from –0.1 nm to +0.2 nm. This value depends on the actual bases involved and so a stretch of DNA may have regions of A, B and C form DNA.

Similarly the helical twist angle is highly dependent on the particular sequence and varies from 27.7° (A_pG) to 40.0° (G_pC). B. DNA is not therefore a smooth, regular double helix but rather shows an irregularity dependent on the base sequence. The alteration of the helical twist angle results, partly from a rolling of adjacent base pairs over one another along their long axes i.e., along a line from C_8 of purine to the C_6 of the pyrimidine.

When the sequence of bases in DNA shows a regular pattern the structure of DNA under goes dramatic changes when there is a region of DNA containing alternating Pu and Py it can lead to wrinkled form of DNA. In more extreme cases to a complete break down of the B form of DNA and conversion to Z-DNA. The wrinkled DNA has deeper minor groves which could lead to changes in the interactions of the duplex with cations and water molecules. It may also affect, DNA protein interaction which could be the reason why the lac repressor binds more strongly to poly [d(A-T)] poly [d (A-T)] than to DNA of general sequence since in the normal recognition site (Lac operator) three fourths of the bases are arranged as alternating Pu and Py. Thus taking up the wrinkled configuration.

If instead of alternating Pu and Py DNA has a Pu-Py dinucleotide followed five bases later by a py-pu dinuctedtide (i.e., PuPy abc PyPu def PuPy ghi PyPu), then the DNA would be bent. In a similar manner runs of As appropriately placed, lead to the formation of a bent molecule.

Under conditions of high salt concentrations DNA with an alternating Pu-Py sequence, ex: poly (d A-T) tends to form a left, handed double helix known as Z-DNA. Which has 12 base pairs per turn in solution. It may also be formed in super coiled DNA molecules where the region of left handed DNA serves to relax the tension of the super coiled molecule.

The secondary structure of DNA is a dynamic one, where in the double helix is not a totally fixed or rigid molecule but under goes considerable internal deformation in a continuous manner. The swinging of small segments of the double helix is referred to as *breathing of the DNA* (in tritium exchange experiments). Sequences which breathe most readily have been calculated, where the opening of a base pair is determined by the stability of the base pair and its nearest neighbour 5′ Pu-Py dinucleotide pairs are stabler than 5′ Py-Pu dinucleotide pair.

Denaturation

When double-Stranded DNA molecules are subjected to extremes of temperature or pH, the Hydrogen bonds of the double helix are ruptured and the two strands separate. The DNA is said to be denatured and forms a random coil. When heat is used as the denaturant the DNA is said to melt and the temperature at which the strands separate is known as the transition temperature or melting temperature (Tm).

The Nitrogenous bases of a poly nucleotides absorb light at 260 nm, but in a double-stranded DNA this absorption is partially suppressed, due to coupling between the transition of neighboring chromophores (i.e., interaction of π electrons of the bases) which is due to stacking of the bases. When the duplex melts, the Hydrogen bonds break and the bases unstack as a result the absorption at 260 nm rises by 20-30%. This is known as *hyperchromic effect* which is used to monitor melting of DNA.

Along with this, the transition is accompanied by a change intensity of the DNA, the single stranded molecule being more dense than the corresponding duplex.

The nature of melting transition is affected by

(a) The (G+C) content of DNA:

Higher the G+C content of DNA more stable will be the molecule and therefore higher will be its melting temperature.

(b) The nature of solvent:

In low concentration of counter ion, denaturation occurs at relatively low temperatures and over a broad range, while at high concentration of counterion the Tm is raised and the transition is sharp.

(c) The nature of DNA:

Most DNA have varying G+C and A+T regions occurring as mosaics, which results in the two strands being held together at G+C regions. This region thus allows the reannealing of the broken H-bond, on lowering the temperature.

Renaturation of DNA:

When 2 DNA strands are returned from the extreme conditions they may reassociate to re-form a double helix. Simple DNA molecules reanneal correctly and instantaneously than complex DNAs, since to reform the double helix they must align themselves perfectly. It depends upon the concentration of DNA and the time allowed. If the solution is quenched at 4° it limits diffusion and prevents the separation of any DNA strand which have mismatched producing solutions of denatured DNA in case of complex molecules.

The complexity of a DNA sample is reflected in the time it takes a solution of DNA of given concentration to reanneal. The reassociation can be followed spectroscopically or by taking advantage of the fact that a duplex DNA binds more strongly to *hydroxy apatite* than a single stranded DNA. Another method makes use of S_1 *nuclease* an enzyme that preferentially digests single stranded DNA.

Buoyant Density of DNA:

Physical properties of DNA are strongly influenced by the percentage of G+C, which also influences the buoyant density of DNA in concentrated CsCl solutions G+C rich DNA has a higher buoyant density than A+T rich DNA.

Isolation and Separation of DNA:

Phenol extraction is used widely in the isolation of both DNA and RNA, which are retained in the aqueous phase while the denatured protein collects at the interface between this and the phenol phase which contains the lipids. RNA can be removed from the DNA by using pure pancreatic ribonuclease or by *isopycnic ultracentrifugation* in a gradient of CsCl. A density gradient established utilising the buoyant densities of the molecules to be separated. Since an equilibrium is established, it is also some times known as *equilibrium ultracentrifugation.*

The buoyant density of double-stranded DNA varies with the G+C content, however RNA has a much higher buoyant density (about 1.9 g/ml) than double stranded DNA (about 1.7 g/ml) that which in turn is higher than of protein (about 1.3 g/ml) single stranded DNA has a slightly higher buoyant density.

Bacterial plasmids can be separated from chromosome DNA using the Isopycnic ultracentrifugation. Chromosomal DNA is linear while plasmids are circular. The difference is exploited by addition of saturating concentrations of intercalating fluorescent dye, *ethidium bromide,* where intercalation requires that the DNA strands be forced apart with a concomitant decrease in buoyant density and partial unwinding of the double helix. The unwinding of the double helix is hindered in the closed – circular plasmid molecule with the result that they bind less ethidium bromide and have a higher buoyant density than linear chromosome.

Chemical Reactions

Some of the important reactions of the DNA are

(a) Strong mineral acids lead to depurination of Nuclei acid which lead to the conversion of sugars into furfural derivatives which give specific colour reactions with orcinol (RNA) or diphenylamine (DNA).

(b) Nitrous acid reacts with amino groups to convert them into hydroxyls. Thus converting

Cystosine ⟶ uracil

adenine ⟶ hypoxanthine

Guanine ⟶ xanthine

Eukaryote DNA

Eukaryotic DNA is contained in a relatively small number of chromosomes. No direct correlation can be made between the amount of DNA in the nucleus and the number of chromosomes in which it is contained. Somatic cell contain diploid number of chromosomes. The characteristic number and morphology of the chromosomes in a particular cell is known as the Karyotype of that cell.

DNA content and chromosome number:

Organism	Haploid DNA picograms	Content Base pairs	Haploid chromosome number
Simian Virus 40	0.000006	5.3×10^3	1
Herpes simplex	0.00017	151×10^3	1
E. Coli	0.005	4.5×10^6	1
Drosophila melanogaster	0.17	0.15×10^9	4
Homo sapiens	2.7	2.4×10^9	23
Xenopus laevis (Toad)	4.2	3.8×10^9	38
Triturus Cristatus (newt)	35	31.5×10^9	12

The amount of DNA is at least an order of magnitude in excess of that required for the known gene coding capacities of the cells where the bulk of the DNA is not expressed.

The minimum size of genome i.e., amount of DNA per cell (C-value) increases with the stage of evolutionary development. However certain amphibia have a 'C' value more than man even more than other amphibia the reason being not clear. This phenomena is known as the *C value paradox* the species with greater amount of DNA should have the advantage of greater cording potential and the disadvantage of the requirement to replicate very large amounts of DNA prior to cell division.

Eukaryote DNA is divided into three frequency classes of *Unique DNA,* moderately repetitive DNA and highly repeated DNA with a considerable overlap between the three categories.

Unique DNA

It comprises about half of the total haploid DNA content sand consists of the Sequences coding for most enzyme functions for which there is only one or a small number of genes per haploid genome. In contrast to the prokaryotic genes which occupy a single uninterrupted sequence of DNA, the majority of Eukaryotic structural genes have extra sequences inserted into the middle of the gene. These interning sequences are called *introns,* which may be small and single as in the case of the gene for tyrosine suppressor tRNA or large and multiple. These intervening regions are transcribed into RNA and then removed in the nucleus to produce the final mRNA. Such interveining sequences appear to be less frequent in the lower Eukaryotes. Majority of structural genes display this phenomenon, and is not universal. The significance of introns is not properly understood and has been suggested to reflect the continual process of evolution within the Eukaryotic chromosome.

Repetitive DNA includes all the rest of the DNA. The sequences represented in this group are generally thought to be those coding for proteins which form major structural components of the cell, ex, genes for rRNA, tRNA, histones etc.

Satellite DNA

It represents highly repeated sequences of which there may be a million or more copies per haploid genome which are usually quite short and are arranged in tandem arrays. The name relates to the method of its isolation on CsCl buoyant density gradient of sheared DNA, where it forms a satellite band separate from the main DNA band due to its differing content of A and T residues.

The distribution of satellite DNA among chromosomes varics some chromosomes have virtually no satellite while other (especially y chromosome) are largely composed of satellite sequences. It appears to be concentrated near the centromere of the chromosome in the *hetero chromatin fraction.* The function of satellite DNA is not known properly. It was thought that satellite DNA was not transcribed since its corresponding RNA was seldom isolated but occasional cases of satellite transcription have been reported. Its transcriptional inactivity lies in with its localization in heterochromatin.

Interspersed Repetitive DNA

It differs from the satellite DNA in that, in it the sequences are not clustered but dispersed singly through out the genome. They are of two classes.

(a) short interspersed nuclear elements (SINES) consisting of repeats shorter than 500 bp and the long interspersed nuclear elements (LINES)

Example of SINES is the Alu family in human cells (it derives its name from the presence of a single site for the restriction enzyme Alu I in the repeat). The Alu sequence is transcribed and it is believed to have become dispersed throughout the genome by reverse transcription and reintegration therefore it is termed as Retroposon.

Fold back DNA or Palindromic DNA these sequences comprise 3–6% of Eukaryotic DNA. The size range is from 300 to 1200 bp.

5′ TGCGATGC : CGATCGCA 3′
3′ ACGCTACG : CGTACGCT 5′

Chromatin Structure

In the interphase the chromosome is referred to as chromatin. Earlier Chromatin was loosely defined and subdivided into two classes – Euchromatin and Heterochromatin. Heterochromatin comprises the dense readily stained areas of the nucleus or chromosome and was thought to represent inactive chromatin. The Euchromatin is more loosely packed, which was thought to represent the transcriptionally active material.

Chromatin consists of DNA, RNA and proteins. In general the amount of protein is equal to or greater than the amount of DNA while the amount of RNA is comparatively small.

The protein content of chromatin can be further subdivided into *histone and nonhistone proteins.*

Histone

The histones are basic proteins of low molecular weight. Each molecule consists of a hydrophobic core region with one or two basic arms. They are classified into five types namely H_1, H_2A, H_2B, H_3 and H_4.

H_1, is very lysine-rich protein of about 216 amino acids which shows a high degree of sequence conservation among Eukaryotes particularly at the central a polar regions. H_2A and H_2B are even more highly conserved and are known as lysine rich histones. The most conserved of all are the Arg-rich histones H_3 and H_4.

The effect of variations in sequence of histones on chromatin structure is not clear. However the histones may be methylated, phosphorylated, etc., and thus affect the interaction of histones with each other or with the DNA.

In sperm cells histones are replaced by other small basic proteins knows as *protamines*.

In the absence of DNA the 'core' histones associate with each other, the predominant species being a homotypic tetramer in the case of H_3 and H_4 a dimer in case of H_2A, H_2B.

Nonhistone Proteins

They occur in approximately equal amounts to histones. Some are the enzymes involved in replication and transcription other resemble the histone in being of low mol. wt. and are known as high mobility group of HMG proteins. They are also basic like the histones and play a structural role, but differ from histones in being only loosely associated with chromatin.

Nucleosome

The DNA of human cell is of the under of 1 m in length and must be condensed in a cell nucleus whose diameter is of the order of 10 μm. Thus packing is essential which should also maintain accessibility during replication. This is possible by a packaging mechanism involving histones and some other chromosomal proteins. The complex of DNA – histone forms the beads called Nucleosomes in which the double-stranded DNA forms the string.

Each Nucleosome is composed of two molecules each of histones H_2A, H_2B, H_3 and H_4 and about 200 bp of DNA. The H_2A, H_2B, H_3 and H_4 molecules form a protein complex called the histone octamer around which the DNA is wrapped. About, 146 bp of DNA are in close contact with the histone octamer to form a nucleosome core particle. The DNA between each core particles is called linker DNA it is about 54 bp long. Histone H_1 is bound to the linker DNA and to the nucleosome core particle. H_1 is responsible for higher order chromatin structures.

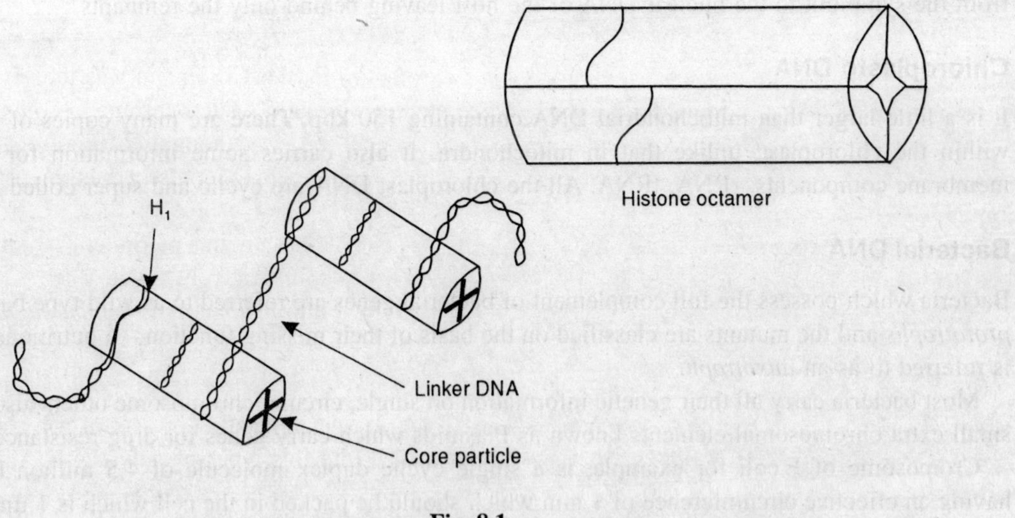

Fig. 8.1

The DNA is thus not only compacted in chromatin it is also rendered partially resistant to nuclease action.

Higher levels of Chromatin Structure

The packaging of DNA into nucleosome accounts for a 10 fold reduction in chromosome chain length. Further reduction comes from higher levels of DNA packaging. The beads on a string structure is itself coiled into a solenoid to yield the 30 nm fiber. It forms when every nucleosome contains a molecule of the H_1. The adjacent molecules of H_1 bind cooperatively bringing the nucleosomes together into a more compact and stable form of chromatin. This achieves a further fourfold reduction in the length.

The 30 nm fibers are finally themselves attached to a non-histone protein scaffold that holds, the fibers in large loops.

There may be as many as 2000 such loops on a large chromosome. Such packaging yields chromosomes that are 5–10 μm in length and 1 μm in diameter. The organization of the DNA into loops accounts for the remaining 200-fold condensation in the length of the DNA.

Extra Nuclear DNA

(a) Mitochondrial DNA

It is found in cyclic, double stranded super coiled molecules with an exception of linear Mitochondrial DNA in paramecium.

Mammalian Mitochondrial DNA are not packaged into nucleosomes. These are about 15 kbp long and can therefore code for 15 to 20 proteins. Yeasts have 75 kbp, plant mitochondrial DNA are still longer.

Mitochondrial DNA can code only for a small proportion of mitochondrial proteins yet it is described as a "lesson in Economy". Introns are absent in mammalian mitochondrial genes but present in some yeast genes. There are genes for rRNAs, tRNAs and some other proteins of the ETC present on the mitrochondrial DNA.

The origin of mitochondria suggests that they rose as symbiotic prokaryotes providing oxidative metabolism to their pre Eukaryotic hosts. Throughout evolution, most of the genes were transferred from the symbiont to the nuclear DNA of the host leaving behind only the remnants.

Chloroplasto DNA

It is a little larger than mitochondrial DNA containing 150 kbp. There are many copies of the DNA within the chloroplast, unlike that in mitochondra. It also carries some information for essential membrane components, rRNA, tRNA. All the chloroplast DNA are cyclic and super coiled.

Bacterial DNA

Bacteria which possess the full complement of bacterial genes are referred to as wild type bacterial or *prototrophs* and the mutants are classified on the basis of their missing function – a nutritional mutant is referred to as an *auxotroph.*

Most bacteria carry all their genetic information on single, circular chromosome others also possess small extra chromosomal elements known as Plasmids which carry genes for drug resistances.

Cromosome of E.coli for example, is a single cyclic duplex molecule of 4.5 million basepairs having an effective circumference of 1 mm which should be packed in the cell which is 1 μm. Thus a complex packaging is required. This is achieved by two main mechanisms.

(a) first the DNA is folded into between 40 and 100 loops,

(b) each of these quasi circles is itself supercoiled independently of others.

Plasmids

They are duplex, super coiled DNA they are stable elements which exist in bacteria and some Eukaryotes. Larger ones are present in only one copy while the smaller may have upto 20 or more copies per cell. These are autonomous, self replicating

REPLICATION OF DNA

Daughter cells produced on cell division contain identical copy of genetic material, which involves the reproduction or replication of DNA.

3 mechanisms have been proposed for the replication of the DNA.

(a) Dispersive, (b) Conservative, (c) Semiconservative.

In the Dispersive mechanism, the two strands do not come apart but act together as a template to form a completely new double helical molecule. In which one daughter molecule would be wholly new and the other totally derived from the parent.

In the semiconservative form it is suggested that the two strands can be untwisted and separated from one another to form two single chains, each base in the single strands can attach to itself the complemently deoxyribonucleotide by the hydrogen – bonding. Thus at the end 2 complete DNA double helices are formed, each with the original molecule and a newly formed DNA molecule.

Experimental results support the semiconservative type.

In a long DNA molecule replication occurs over one short stretch at a time and the two parental strands, separate only at the point of replication to produce a Y-Shaped molecule as the replication fork passes along the DNA. Such replication forks appear as bubbles in the replications DNA molecules. One daughter DNA chain grows in the 5′–3′ and not from 3′–5′ by the production of *okazaki fragments.*

One side of a replication fork the DNA is single stranded (supporting the discontinuous mechanism). The strand of DNA which is made continuously in the direction of fork movement is called the leading strand while that made discontinuously in the reverse mode is the lagging strand.

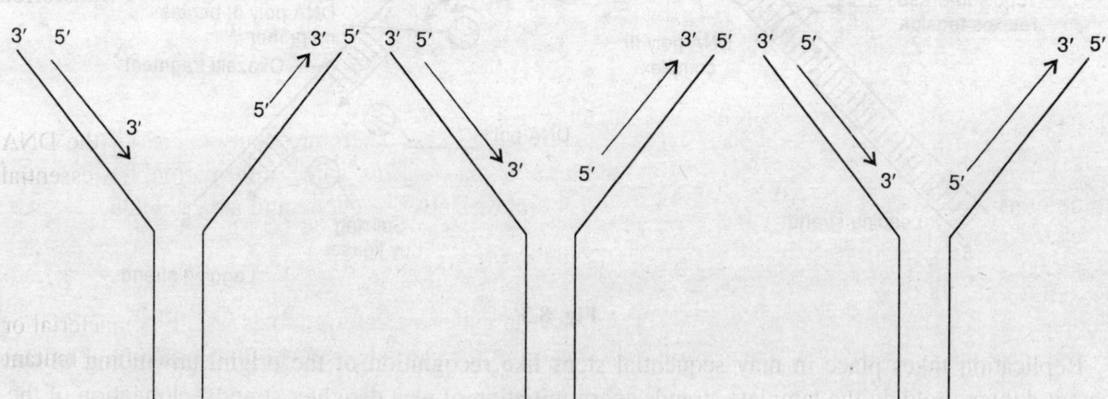

Fig. 8.2

Continuous Synthesis

Discontinuous synthesis of one strand of DNA was postulated in order to over come certain problems but no problems appear to exist, with the chain growing 5′–3′ direction. Thus there is no priori reason why this chain should not be made continuously. Infact this would be the case when DNA synthesis is proceeding rapidly.

The Process

DNA synthesis occurs at a replication fork which progresses along the DNA molecule. Many proteins are required for the same, namely.

1. DNA polymerase – required to add dNTPS to the growing leading strand and to the growing okazaki pieces on the lagging strand.
2. A primase to initiate the okazaki pieces.
3. An exonuclease to remove the primer.
4. A ligase to join the okazaki pieces.
5. An unwinding protein required to unwind the duplex DNA at the replication fork.
6. A Topoisomerase to relax the tension caused by unwinding duplex.
7. A binding protein to stabilize single stranded DNA exposed to progression of the leading strand prior to initiation of okazaki pieces.
8. A Gyrase to help unwind the double helix ahead of the replication fork or for initiation of replication.

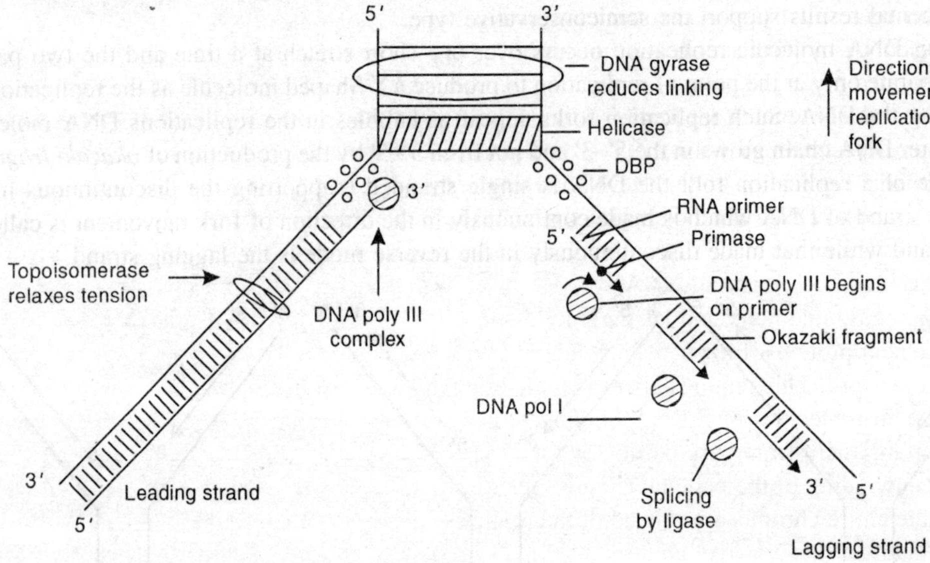

Fig. 8.3

Replication takes place in may sequential steps like recognition of the origin, unwinding of the parent duplex, holding the template strands apart, initiation of new daughter strands, elongation of the daughter, rewinding and termination of the replication. All of these proceed at a very high rate and it is extremely accurate. The entire complex of 20 or more replication enzymes and factors is called as DNA replicase system or Replisome.

In the E.coli, three different DNA polymerase (polymerase/Pol) I, II and III occur in the cell. Pol I is the most abundant one. The enzyme is able to synthesize DNA from fours precursor molecule namely of dATP, dGTP, dCTP and dTTP provided a template is given. Neither 5′ monophosphate nor 5′– diphosphates, or 3′ (mono, di and tri) phosphates can be polymerized. Pol III is responsible for the elongation of DNA. It contains Zn^{2+} and requires Mg^{2+} and functions similarly to Pol I.

Both Pol I and III have three enzymatic activities. In addition to polymerase activity these DNA polymerases also act as 5′–3′ exonucleases and as 3′–5′ exonucleases, the function of Pol II is still not known.

DNA polymerases			
Activity catalysed	I	II	III
5′ – 3′ polymerizations	✓	✓	✓
5′ – 3′ exonuclease	✓	—	✓
3′ – 5′ exonuclease	✓	✓	✓
Mol. Wt.	109,000	120,000	400,000
Molecules percent	400		10

One strand of DNA is replicated continuously in the 5′–3′ direction, the other strand is made discontinuously is short pieces called okazaki fragments. Formation of okazaki fragments require as a primer a short length of RNA, complementary to the DNA template strand.

The RNA is made in 5′–3′ direction from ATP UTP CTP and GTP precursors Primase. To the 3′ end of this short RNA primer are added deoxy ribonucleotide units complementary to the template DNA. The action of pol III makes possible the addition of 1000 to 2000 deoxyribonucleotide. The RNA primer is then removed by 5′–3′ exonuclease activity of pol I and is replaced by complementary deoxyribonucleotide by pol I acting in its polymerase mode and using 3′ end of the preceding okazaki fragment as a prirrer.

The new okazaki fragments are joined to the lagging strand by the enzyme DNA ligase which can form a phospho diester bond between 3′ – OH group and 5′ phosphate group of the newly made okazaki fragment. Formation of this new bond requires energy which is provided by the coupled hydrolysis of pyrophosphate bond of NAD+ (in bactena) or ATP (in animal cells). The action of DNA ligase is most efficient when two DNA fragments to be joined are completely base paired with the complementary template strand of DNA.

Unwinding the double helix and keeping the 2 strands apart to allow replication made possible by several specialized proteins. Enzymes known as helicases unwind short segments of DNA just ahead of the replication fork. This requires energy yielded by ATP. After a short sequence of DNA has been unwound, several molecules of a DNA binding protein (DBP) bind tightly to each of the separated strands preventing them from base pairing again.

The rapid unwinding of the parental strands creates another problem. Without some special swivel′ mechanism, the entire chromosome ahead of the replication fork would be forced to rotate at this high speed. To counteract this a swivel is introduced by a transient break in one strand of the DNA which is very quickly and accurately respliced after one or more revolutions. The transient break and splice is introduced by enzymes known as Topoisomerases. In prokaryotes the *Topoisomerases* are called as *DNA gyrase,* which not only permits swiveling of the DNA but actively twists it in the direction which favours unwinding of the template strands at the fork.

Thus gyrase helps helicase to unwind the DNA for replication. Its activity also results in super coiling of the chromosome and is responsible for keeping DNAs of bacterial cells in a super coiled form.

Source of precursors

There are 2 distinct pathways for the synthesis of the DNA precursors the de novo path way and the salvage pathway.

In the denovo pathway, ribonucleoside monophosphates are synthesized from phosphoribosyl, pyrophosphate, amino acids, CO_2 and NH_3. The nucleoside monophosphates (NMP) are then converted to NDP by enzymes called kinases. These kinases are specific for each base but nonspecific with respect to ribose or deoxyribose.

The NDP are then converted to their deoxy form by an enzyme, ribonucleotide reductase, it is not base specific.

The enzyme Nucleoside diphosphate kinase, which is neither base specific nor sugar specific then forms the triphosphate.

Due to this lack of base specificity of both nucleoside disphosphate kinase and ribonucleotide reductase, the nucleotide *is synthesized, due to which considerable effort* is devoted to prevent its incorporation.

Synthesis of TTP is unique since TMP is formed by methylation of dUMP, reaction beings catalysed by Thymidylate synthetase and the methyl group by which uracil and thymine differ is obtained from 5, 10 – methylene tetrahedra folate.

Inhibitors of tetrahydrofolate synthesis makes it essential for exogenous thymine or thymidine without which the cells die. A drug that inhibits the conversion of dihydrofolate to tetrahydofolate namely Metho trexate is thus commonly used in cancer chemotherapy.

In the salvage pathway, free bases and nucleosides obtained either by degradation of nucleic acids or form growth medium are built upto nucleoside monophosphate.

DNA Polymerases

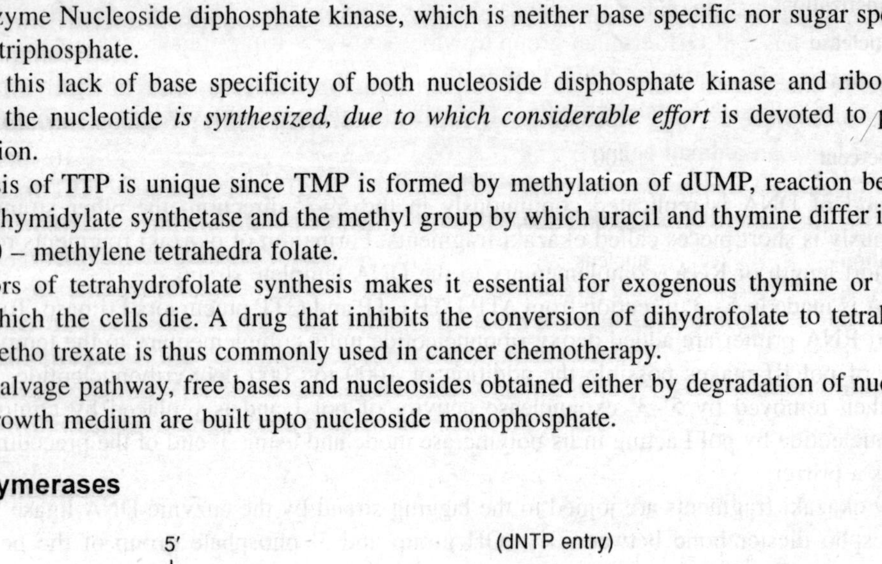

Fig. 8.4

DNA polymerase 1 has a Mol. wt. 109,000 about amino acid long with a diameter of 65 A0. The enzyme binds with DNA only if it has a partially single stranded section. Within the enzyme are a number of essential sites that enable it to function (diagram). At one site a section of template DNA is held in place, and adjacent to this is a site holding a sequence of nucleotides complementary to the template called primer. At the growing tip of the primer the enzyme bears a site, that can match an incoming nucleotides triphosphate to a complementary nucleotide of the DNA template. If the match is successful, then the incoming nucleotide is chemically bound via a phosphate group to the 3′ position of the primer.

If a mismatched nucleotide occupies the primer site, the polymerase enzyme has exonuclease activity which will remove this incorrect nucleotide in the 3′–5′ direction and replace it with the proper nucleotide.

In addition the 5′–3′ exonuclease activity allows the enzyme to degrade nucleotides that may have been previously incorporated, but mismatched, such as ribonucleotides.

The primer terminus site on the enzyme cannot be occupied unless the primer carries at this position a nucleotide that has a 3′ OH terminal group to which only a 5′ triphosphate monomer can be added. Thus this enzyme can synthesize DNA only in the 5′–3′ direction. The exonuclease editing function also explains why the DNA chain growth is restricted to only one direction (5′–3′).

Eukaryote (Metazoan animall cell)

	α	β	γ
Mol. wt.	120,000–220,000	30,000 – 50,000	150,000 – 300,000
Location	nucleus	nucleus	Mitrochondria
Function	replication of nuclear DNA	repair	replication of mitrochondrial DNA
Ability use RNA	yes	no	no
Primer Exonuclease 3′–5′ activity 5′ – 3′	no	no	uncertain
Molecules per cell	20,000 – 60,000	60,000	?

Replication of Circular DNA Molecule

The first demonstration that E. Coli DNA replicate as a circle came from auto radio graphic experiment. A replicating circle is schematically like θ, hence it is usually called θ replication.

In rolling circle replication model, a nick in one of the strands allows the 5′ end to become attached to a membrane while the other end is free to act as a primer so that DNA synthesis forms a complement of the unbroken parental strand. As synthesis proceeds the unbroken parental DNA rolls and unwinds thus extending the template upon which new DNA is synthesized. Meanwhile the tail strand attached to the membrane is also lengthened and provides a template upon which a complementary DNA can be synthesized.

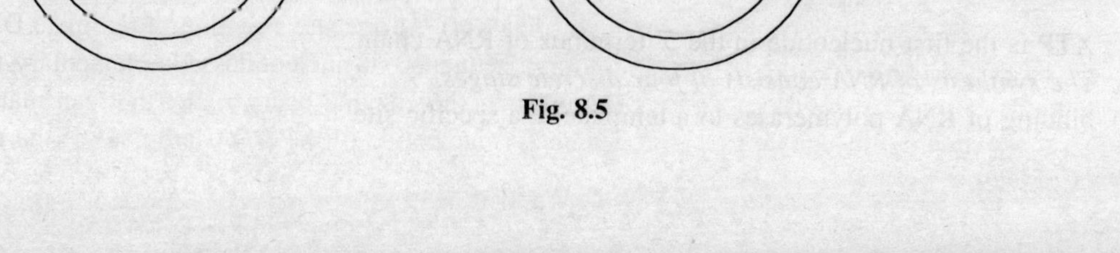

Fig. 8.5

Initiation of synthesis of the leading strand. All known double stranded DNA initiate a round of replication at a unique base sequences, called the replication origin or ori. The sequence is specific to each organism. Initiation can occur in 2 ways – de novo initiation in which the leading strand is started afresh and covalent extension, in which the leading strand is covalently attached to a parental strand.

Transcription

Expression of the gene occurs by the transfer of genetic information from DNA to RNA and then to the protein molecules.

The process where by RNA molecules are initiated, elongated and terminated is called transcription where one strand of DNA acts as a template and the polymerization is catalysed by RNA polymerase.

The essential features of RNA synthesis are

(a) The precursors in the synthesis of RNA are the four ribonucleoside 5′ – phosphates (NTP).
(b) In the polymerization reaction a 3′ – OH group one nucleotide reacts with the 5′ triphosphate of a second nucleotide pyrophosphate is removed and a phosphodiester bond results.
(c) The sequence of bases in an RNA molecule is determined by the base sequence of the DNA.
(d) In any particular region of DNA only one strand serves as a template.
(e) The RNA chain grows in the 5′–3′ direction i.e., nucleotides are added only to the 3′ OH end of the growing chain.

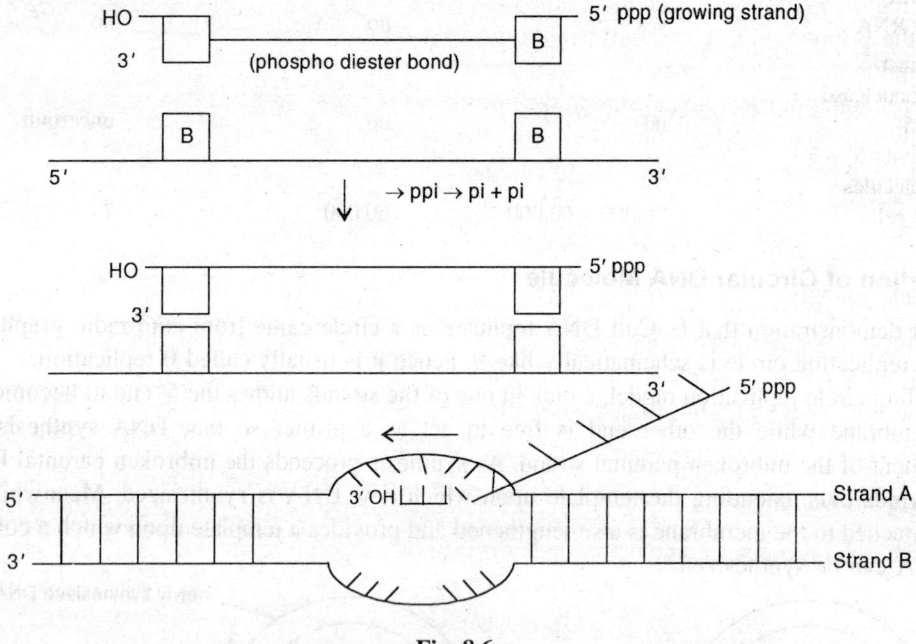

Fig. 8.6

(f) RNA polymerases are able to initiate chain growth i.e., no primer is needed.
(g) Only ribonucleoside 5′ triphosphates participates in RNA synthesis,

$$nNTP + XTP \xrightarrow[Mg^{2+}]{DNA,RNA\text{-}P} (NMP)_n - XTP + nPPi$$

XTP is the first nucleotide in the 5′ terminus of RNA chain.

The synthesis of RNA consists of four discrete stages.

(1) binding of RNA polymerases to a template at a specific site

(2) initiation
(3) chain elongation
(4) chain termination are release.

The structure of RNA polymerase helps in understanding better these processes, E.Coli. RNA polymerase. It consists of 5 subunits, 2 identical α subunits and one each of β, β' and σ.

It has a mol. wt. of 465,000. The σ subunit dissociates from the enzyme during the elongation stage. The core enzyme consists of α_2, β, β' while the holo enzyme consists of α_2, β, β' and σ.

RNA polymerase is sufficiently large to accommodate 41 to 44 base pairs.

Site selection

The Promoter

The first step in transcription is the binding of RNA polymerase to a DNA molecule. It occurs in particular regions called promoters which are sequences in which several interactions occur. RNA polymerase must recognize a specific DNA sequence, attach in a proper conformation, locally open the DNA strands in order to gain access to the bases to be copied, and then initiate synthesis. These events are guided by the base sequence of the DNA, the polymerase σ subunit and also in a few cases by auxiliary protein.

In a region 5–10 bases to the left of the first base copied into mRNA is the right end of a sequence called the pribnow box (commonly called 'box') to indicate a sequence that occurs repeatedly as a transcription or regulatory signal.

After binding of the holo enzyme to the pribnow box, the DNA helix in an open promoter complex is locally unwound starting from 10 bp from the left end of the pribnow box and extending about 20 bp past the position of the first transcribed box.

The promoter strength determines the number of copies of each protein molecules present in the cell.

Auxiliary Proteins

Some bacterial promoters require an activator protein for effective initiation, ex : cAMP receptor protein, CRP on Lac promoter. In the absence of CRP, RNA polymerase holo enzyme binds to the pribnow box, but a closed-promoter complex results in which no dissociation of the DNA strand occurs.

RNA Chain Initiation

Once the open-promoter complex has been formed, RNA polymerase initiates synthesis. It contains 2 nucleotide binding sites called initiation site and elongation site. The initiation site bind purine triphosphate ATP and GTP and ATP is usually the first nucleotide in the chain. Thus the first DNA base that is transcribed usually is thymine. The initiating nucleoside triphosphate forms a 'H' bond with the complementary base of DNA.

The elongation site/catalytic site is then filled with an NTP that has the ability to form 'H' bond with the next, base of DNA strand. The 2 nucleotides are joined together, the first base is released from the initiation site and thus initiation is complete.

DNA template and RNA polymerase move relative to each other such that to catalytic site and binding site shift by one nucleotide.

The drug Rifampicin binds to the β subunit of RNA polymerase blocking the transition from chain initiation to elongation phase. It is an inhibitor of chain initiation but not of elongation.

Chain elongation

After several NTPs, are added to the growing chain, RNA polymerase under goes a conformational change and loses σ subunit. This marks the transition from the initiation phase to the stable forward movement of the elongation phase and elongation is carried out by the core enzymes. The core enzyme moves along the DNA, binding a nucleoside triphosphate that can pair with the next DNA base and opening the DNA helix as it moves. The DNA recloses as synthesis proceeds.

The chain elongation does not occur at a constant rate, i.e., synthesis slows down when particular regions of DNA are passed. The reduction in rate is called a 'pause' which accelerates further and the process is repeated.

Termination and release

It occurs at specific base sequences within the DNA molecule. They are of 2 types.

1. simple terminators, and
2. those that require auxiliary terminating factors.

In the factor independent terminating sequences three important regions have been identified.

(1) There is an inverted repeat containing a central non repeating sequence in DNA strand, which is capable of intrastrand base pairing, forming a stem and loop in the RNA molecule. This could serve a role in termination and also renders the RNA resistant to degradation by RNA phase II.

(2) Near the loop end there is a high G+C content. RNA polymerase usually slows down when synthesizing the corresponding RNA segment.

(3) This followed by a sequence A-T pairs.

The second class of termination requires proteins called Rho, which is protein that binds tightly to RNA. When bound segments rich in 'C', it acquires a powerful ATP – clearing activity, that is essential for its action in termination.

Transcription in Eukaryotes

The basic features are similar to bacteria, but there are five notable differences.

(a) Eukaryotic cells contain 3 classes of nuclear RNA polymerase which are responsible for the synthesis of all RNA.

(b) Both the 5′ and 3′ termini are modified, a cap is found at 5′ end of all mRNA and a long poly (A) is found at 3′ end of most mRNA.

(c) The mRNA molecule which is used as template for Protein is usually processed.

(d) Eukaryotic mRNA molecules are mono cistronic.

(e) Many mRNA molecules are long lived.

RNA Polymerase

Three different, RNA polymerase I, II, III are known is eukaryotes, each responsible for the synthesis of particular class of RNA.

RNA pol I makes only rRNA

RNA pol II all mRNA

RNA pol III tRNA 5 sRNA of ribosomes.

Each is a high mol. wt. (500,000) protein consisting of 2 large subunits and up to 10 small subunits.

The biochemical reaction catalysed by the Eukaryotic RNA polymerases are same to that of E.coli. RNA polymerase.

Promoters of RNA polymerase II

(1) There is a universal sequence about 25 bp upstream from the transcription start site consisting of TATAAAT. It is known as TATA or Hogness box and is similar in sequence to the pribnow box.
(2) Many, but not all have another common sequence in the 75 region in which T and C are, frequent. This is known as the CAAT box.
(3) The TATA box and other Sequences if present, are sites of binding of transcription factor, but not of RNA polymerase.
(4) RNA pol II does not interact directly with any part of the promoter and is dependent on the presence of transcription factors.

Promoter of RNA polymerase III

Pol. III is responsible for synthesis of 5 sRNA and tRNA. Its promoter differ significantly from pol. II as they are down stream frosm the transcription start site and within the transcribed DNA. This indicates that RNA pol. II reaches forward to find the start point RNA pol III reaches backward.

Classes of RNA molecules (prokaryotes) 3 classes of RNA – rRNA, tRNA, mRNA in Eukayotes sRNA. All are transcribed from DNA base sequences sand significant differences exist between the structures and modes of synthesis of RNA molecules of Prokaryotes and Eukayotes.

mRNA

Amino acid sequence is procured from the DNA by the mRNA, the nucleotide sequence of mRNA is then read in groups of 3 bases (codon) from a start codon to a stop codon where each codon corresponds to one amino acid or a stop signal.

DNA segment corresponding to one polypeptide chain plus the translational start and stop signals for protein synthesis is called a cistron and an mRNA coding for a single polypeptide is called mono cistronic mRNA. In prokaryotes mRNA are polycistronic. The segment of RNA corresponding to a DNA cistron is often called a Reading frame since it is read by the protein synthesising system. Cistrons contained in polycistronic mRNA often correspond to proteins of a single metabolic pathway. In a way this helps to regulate synthesis of a particular protein with the same signal.

Size of mRNA varies smallest being 150 to 8000 nucleotidase. A typical mRNA contains 3000–8000 nucleotides. In addition to reading frame, start and stop sequences for translation other regions are also important, Ex. translation of an mRNA molecule seldom starts exactly at one end of the RNA and proceeds to the other end, instead, initiation of synthesis of the 1st poly peptide of a poly cistronic mRNA may begin hundreds of nucleotides from the 5′–P terminus of the RNA. The section of nontranslated RNA before the coding region is called a leader. Polycistronic mRNA typically contain inter cistronic sequences usually tens of bases long. Life time of a prokaryotic mRNA is short and 1/2 life of a typical mRNA is a few minutes. This feature has an important regulatory function. The short life time of bacterial mRNA is one criterion used to identify mRNA in prokaryotes – pulse chase experiments.

rRNA and tRNA

Proteins are synthesized on the surface of an RNA containing particle called a ribosome, these particles consist of several classes of rRNA which are stable molecules and have various function.

Amino acids do not line up against the mRNA template independently during protein synthesis but are aligned by means of a set of adapter RNA called tRNA, each capable of reading 3 adjacent mRNA bases and placing the corresponding amino acid on the ribosome at which a peptide bond is formed.

The following 3 properties of these molecules indicate that they are not the immediate product of transcription

(a) All are terminated by a 5′ NMP and not 5 NTP.

(b) Both rRNA and tRNA are much smaller then the primary transcript.

(c) All tRNA molecules contain bases other than A, G, C and U not present in the original transcript.

Thus these molecular changes are made after transcription by post transcriptional modification/ processing.

Processing tRNA

(1) Formation of 3′ OH terminus

It involves the action of an endonuclease that recognizes a hair pin loop and an exonuclease that recognizes the 3 base sequence CCA. After endonuclease digestion at site 1, the 7 bases upstream are removed by an exonuclease called RN ase D.

(2) Formation of the 5′P terminus

It is formed by an enzyme called RNase P. It does not recognize a specific base sequence at the cleavages site or any where else, but responds to the over all 3-D conformation of the tRNA molecule with its several hair pin loops and then makes a cut at the right place.

(3) Production of modified bases

The final modification is to produce altered bases of tRNA. Two uridines are converted to seudouridine (ψ), 2 uridine to, two 4-thiouridines, one Guanosin to 2′-0-Methyl, Guanosine and one Adenosine to isopentenyl adenosine. These modified bases occur in most tRNA.

When the +RNA molecule lacks the 3′ CCA terminal it is added by the enzyme tRNA nucleotidyl transferase.

Processing rRNA

Bacterial ribosomes contain 3 kinds of rRNA. 5s rRNA, 16s rRNA and 23s sRNA and contain 120, 1541 and 2904 nucleotides respectively, these are cleared from a continuous transcript having more than 5000 nucleotides.

A rRNA transcript is usually cut while it is being synthesized by several enzymes acting in sequences. The first cuts are usually made by RNase III.

Eukaroytic mRNA

The 5′ terminus of Eukaryotic mRNA molecule is not a 5′ triphosphate group. It is altered by the formation 5′–5′ linkage of methylated Guanosine derivative, 7-methyl Guanosine (7-MeG).

The first step is the reaction between GTP and the terminal triphosphate of the RNA.

$$Gpp + pppApNpNp \longrightarrow GpppApNpNp + pp + pi$$

by the enzyme *Guanylyl transferase.*

This is followed by the methylation of the initial terminal G/A to form 2′-0-Methyl Guanosine (2′-0-MeG) or 2-0-Methyl Adenosine (2′-0-MeA).

Thus the structure of the resulting unit (7-MeG)-5′-ppp-5′-[2′-0Me (G/A)]-3′-p-5′-nucleoside-3′-p. This region is called as a *cap*. In yeast and slime moulds the most abundant cap structure is cap O in which 7 - MeG is added to the 5′ terminus and no further Methylation occurs. This structure is not found is animal cells.

In viral mRNA and most animal cell mRNA methylation of ribose also occurs on the second nucleotide giving the structure of cap 1.

In some cases Methylation also occurs in the third nucleotide - Cap 2.

Capping occurs shortly after initiation of synthesis of mRNA and it is believed that it is required for efficient protein synthesis. It may also function to protect the mRNA from degradation by nucleases.

Most animal mRNA are terminated at the 3′ end with a poly (A) 20–200 nucleotides long. It is synthesized by the nuclear enzyme poly (A) polymerase.

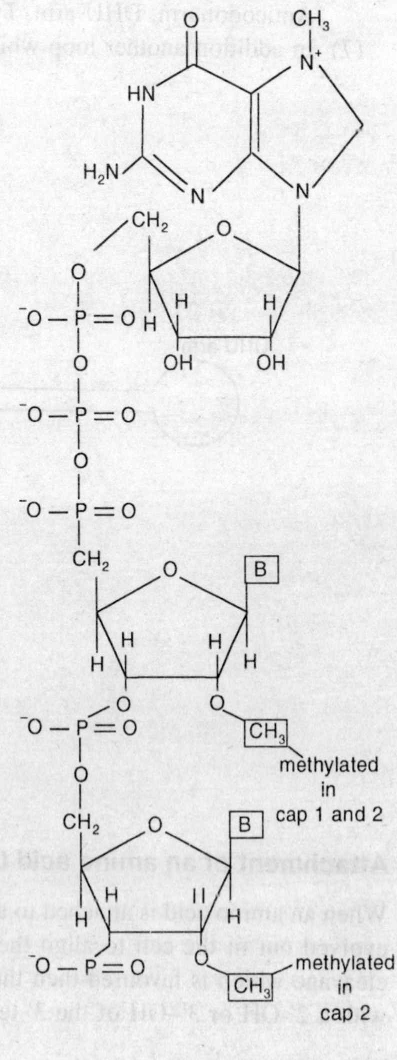

RNA splicing

A characteristic of most of the primary transcripts of higher Eukaryote if the presence of untranslated intervening sequences (introns), which are excised from the primary RNA transcript. The amount of discarded RNA ranges from 50 to nearly 90% of the primary transcript.

The Decoding System – tRNA and the Aminoacyl synthetases

The relation between the base sequence of the coding region of mRNA and the aminoacid sequence of the protein translated is provided by the Genetic code. But, the amino acid do not line up themselves along a mRNA molecule. The decoding system consists of two different, molecules –

(1) small adaptor RNA molecules called tRNA and
(2) the enzymes aminoacyl synthetases.

There are many different tRNA molecules in a particular cell. *Each having 3 important regions.*

(a) A contiguous sequence of 3 bases that can Hydrogen bond by base pairing to each codon this sequence is called as the anticodon.
(b) The amino acid attachment site. The aminoacyl synthetase match the amino acid and the anticodon correctly for which purpose individual tRNA should be recognized and therefore tRNAs have a Recognition region.

A tRNA molecule and its corresponding aminoacyl synthetase are called *cognates*.

All tRNA are small, single stranded nucleic acids ranging in size from 73 to 93 nucleotides. By careful comparison of the sequences in many organisms the idea of a 'consensus' that tRNA consists of 76 nucleotides arranged in a clover leaf from has been proposed. The nucleotides are numbered 1 to 76 from the 5′–p terminus.

Standard tRNA has the following features:

(1) Bases in positions 8, 11, 14, 15, 18, 19, 21, 24, 32, 33, 37, 48, 53, 54, 55, 57, 58, 60, 61, 74, 75, 76 are invariant.

(2) The 5′–p –terminus is always base paired, which contributed to its stability.

(3) The 3′ OH is always a 4 base single stranded region having the sequence XCCA-3′-OH in which X is any base. This is called as the CCA/acceptor end. The adenine in the CCA is the site of attachment of the amino acid by the cognate synthetase.

(4) There are many modified bases ex. Dihydrouridine (DHU), ribosyl thymine (rT) pseudouridine (ψ) and inosine (I).

(5) There are 3 large single stranded loops the lower most or anticodon loop contains 7 bases. The anticodon occupies position 34, 35 and 36. It is always preceded by 2 pyrimidines and followed by a modified purine. There is the DHU loop and the TψC loop.

(6) There are four double stranded region called stems or arms and are called accordingly as Anticodonarm, DHU arm, TψC arm and the acceptor arm.

(7) In addition another loop which is highly variable called extra arm is also present.

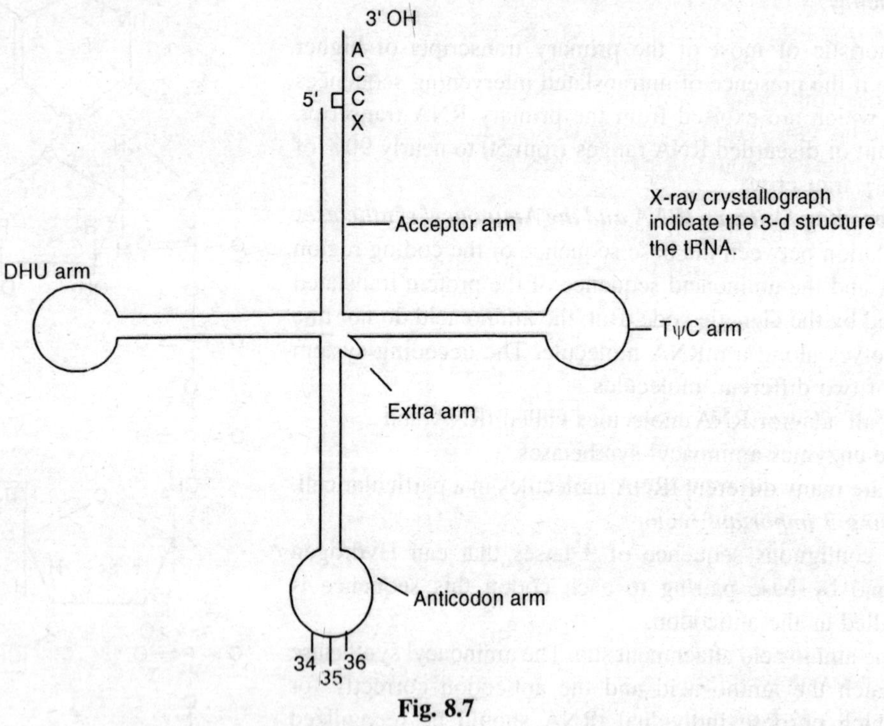

Fig. 8.7

Attachment of an amino acid to a tRNA

When an amino acid is attached to a tRNA, the tRNA is said to be amino acylated or charged. This has evolved out in the cell to align the amino acids according to a certain order and the peptides bond cleavage which is favoured then the bond formation is reversed by first forming an amino acid ester with a 2′-OH or 3′–OH of the 3′ terminal nucleotide of tRNA.

Acylation is accomplished in two steps both of which is catalysed by aminoacyl synthetase. In the 1st or activation step an amino acyl AMP is synthesised from an amino acid and ATP.

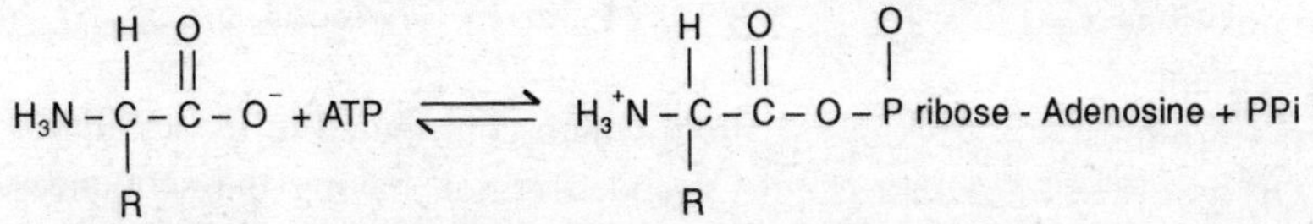

In the second or transfer step the amino acyl-AMP, which is bound by the enzyme, reacts with tRNA to form acylated tRNA and AMP.

$$\text{Aminoacyl - AMP - tRNA} \rightleftharpoons \text{Aminoacyl - tRNA + AMP}$$

The hydrolysis of Pyrophosphate drives the reaction in the forward direction.

Aminoacyl synthetases

Each amino acid has at least one and usually one aminoacyl synthetase. Amino acids specified by more than one codon have more than one synthetase.

There are four classes of synthetases.

(a) Those consisting of a single poly peptide chain.
(b) Those having two identical chains.
(c) Those containing four chains of 2 types (copies each)
(d) One enzyme with four identical subunits.

Each synthetase should be able to recognize its cognate-tRNA molecule, which spans the entire molecule. Results of Photo activated cross linkage and RNase protection indicate that there are several points of contact primarily in the DHU, the acceptor stem and the anticodon loop. The CCA terminus is positioned at the active site of the enzyme. The contact point are all located on the same side of the 3-d structure.

EXERCISE:

1. Explain in detail physical and chemical structure of DNA.
2. Explain different conformational forms of DNA.
3. Explain in detail replication of DNA.
4. Write short notes on
   (a) Unique DNA.
   (b) Satellite DNA.
   (c) Interspersed repetitive DNA.
   (d) Extra nuclear DNA.
   (e) DNA Polymerases.
   (f) RNA Polymerase.
   (g) mRNA.
   (h) rRNA and tRNA.

9

Mendelism

HISTORICAL PERSPECTIVE

In common striking similarities exist in the appearance of parents and their offsprings in any species. This is due to transmission of genetic traits from one generation to the next. According to the ancient literature, Manu Smriti of 700–500 B.C., it is believed that "Subcejan Sukshetre Jayate Sampadyate". It means that a good seed and good field yield abundant produce, highlighting the importance of not only the genetic make up but also the influence of environment on the proper expression of the genetic potential. Plato and Aristotle, Greek philosophers during (427–347 B.C.), have also recognised the role of heredity in human beings. William Harve in 1516 has postulated that all living organisms developed from an egg cell. Thomas Fairchild in 1717 has produced the first hybrid by crossing Dianthus barbetus with Dianthus caryophyllis. Joseph Kolreuter in 1763 has recognised the role of both parents in heredity and performed back crosses. Lamarck in 1809 proposed the theory of inheritance of acquired characteristics, for example, the long neck of giraffe. Charles Darwin in 1859 has published his famous theory of evolution by natural selection.

Though it was not easy to determine the exact nature of the genetic material for a long time. Until mid–eighteenth century many biologists in Europe believed that small organisms could arise all of a sudden from decaying matter spontaneously. (The theory of spontaneous generation). However, this theory was finally discarded by Pasteur during 1822–1895 and Tyndall during 1820–1893; they have proved that observed growth was due to contamination of living spores.

The modern science of genetics started with the discovery of the laws of inheritance by Gregor Mendel. Mendel published his research paper on "Experiments in Plant Hybridization in 1866; which is the foundation of laws of heredity (Mendel's laws). Although his work was ignored for a longtime, in 1900, de Vries in Holland, Tschermark in Austria and Correns in Germany have independently rediscovered the Mendel's laws of heredity.

TERMINOLOGY IN GENETICS

1. **Gene:** A gene is defined as a short sequence of nucleotides on DNA, responsible for the expression of a particular character. It is usually represented by a symbol - a capital letter for dominant gene and a small letter for recessive gene. The position where a gene is present on DNA is called as 'Locus'.

Ex : The gene which is responsible for tallness of plant is represented as 'T' and the gene responsible for dwarfness is represented as 't'.

2. **Allele or Allelomorph:** It is an alternative form of gene controlling the same primary character but a different manifestation. It usually arises as a result of mutation of the gene. It occupies the same locus on the homologous chromosome.

Ex: Round and wrinkled shapes of the seed are different forms of shape of the seed. Round shape is governed by gene 'R' while wrinkled shape by its allele 'r'.

3. **Dominant and Recessive:** The character which is expressed in heterozygote is called as dominant character and the character which is suppressed in a heterozygote is called as recessive character.

   The allele which is expressed in heterozygote is called as dominant allele and the allele that is not expressed in heterozygote is called as recessive allele.

4. **Homozygote and heterozygote:** If the two alleles governing a particular character are similar, it is called as homozygote while if the two alleles governing a particular character are dissimilar it is called as heterozygote. In a heterozygote, out of the two different alleles only one of them is generally expressed.

Ex: The two alleles of a homozygous tall individual is 'TT'. The alleles of a heterozygous tall individual is 'Tt'.

5. **Phenotype and Genotype:** (Terms introduced by Johannsen) Phenotype is the external appearance of an organism and Genotype is the genetic constitution of an organism. Two individuals having the same phenotype may differ in their genotype. i.e., a tall plant may either be homozygous tall (TT) or heterozygous tall (Tt).

Ex: Round shape of the seeds, tallness, red and green colours are examples of phenotypes. The Genotype of Round seed is RR or Rr. and tall is 'TT' or 'Tt'.

Character	*Phenotype*	*Genotype*
Length	(a) Tall	TT/Tt
	(b) Dwarf	tt
Color of Cotyledons	(a) Red	RR/Rr
	(b) Green	rr
Shape of the seed	(a) Round	RR/Rr
	(b) Wrinkled	rr

6. **Pureline:** Homozygous individuals when selfed always produce homozygous individuals of the same genotype. Generations of such homozygous individuals which produce offsprings of only one type form a pureline.
7. **Reciprocal cross:** It is a cross concerning the same characteristics but with reversed sexes.

For example, if a black male guinea pig is crossed with a white female one, the reciprocal cross of this cross is a black female one crossed with a white male one.

Black ♂ × white ♀
Black ♀ × white ♂ } are Reciprocal crosses

MENDEL'S WORKS

1. Gregor Johann Mendel was a monk in a monastery at Brunn, Austria.
2. After seven year of work on pea plants, he presented the results which are now known as

Mendel's Laws in 1865. But his research work did not receive due recognition and was neglected until 1900.

3. In 1900, three scientists - Devries in Holland, Correns in Germany and Tschermak in Austria, working independently rediscovered Mendel's Laws, which laid the foundation for the advancement of Genetics.
4. Mendel performed several experiments on Pisum sativum. His success is mainly due to (a) Pea plants can be easily grown and complete their life cycle within a few months so that several successive generations could be studied. (b) He studied one trait at a time. After establishing the behaviour of a single trait, he studies two traits together. (c) He performed several identical matings so that large number of progeny can be studied. (d) He selected seven pairs of traits in pea plant. In all these pairs, one trait is dominant and the other one is recessive. The seven pairs of traits are (i) Tall and dwarf plant (ii) yellow and green seeds (iii) Smooth and wrinkled seeds (iv) Red and White flowers (v) Axillary and terminal flowers (vi) (Green and yellow pods and (vii) Inflated and constricted pods. (e) The genes for the seven pairs of traits are on different pairs of chromosomes and therefore they assorted off independently.
5. **Mendel's Laws:** From the results of Mendel's experiments, he proposed certain laws which are called as Mendel's Laws of Inheritance. Mendel's Laws are (a) Law of Segregation and (b) Law of Independent assortment.
6. Law of Segregation states that when a pair of contrasting alleles are brought together in a heterozygote, the two alleles remain together without being contaminated and when gametes are formed from the hybrid, the two alleles separate out from each other and one enters into each gamete. OR The separation of paired alleles from one another in F_1, and their distribution to different gametes so that each gamete receives only one allele is called as Law of segregation or Law of purity of gametes.
7. Law of independent assortment states that if the inheritance of more than one pair of characters or traits is studied simultaneously, the genes for each pair of character which are brought together in F_1 individual, assort out independent of the other pair during gamete formation and one has no influence on the other.
8. Law of segregation can be best explained by Monohybrid cross while Law of independent assortment can be explained by Dihybrid cross.
9. Monohybrid cross is a cross between two individuals which differ in one contrasting character.
10. Mendel, while studying the inheritance of one character at a time, selected a pure/homozygous tall plant which was about 6–7 feet in height and a dwarf plant which was 1½ feet in height.
11. Tallness is governed by the gene 'T' and dwarfness of a plant by its allele 't'. Hence, the genotype of pure tall plant can be represented as TT and the genotype of a dwarf plant can be represented as tt.
12. A pure tall plant was crossed with a dwarf plant i.e., affected cross fertilization between these two. The gametes formed from tall plant carry the gene 'T' while the gametes formed from dwarf plant carry the allele 't'. These gametes fuse at random and the offspring produced from the seeds of such a cross were all tall plants having a genotype 'Tt'. These heterozygous tall plants were as tall as the homozygous individuals indicating that the gene T is dominant over its allele 't'.
13. When these F_2 progeny were selfed, each F_1 hybrid during gamete formation forms two types of gametes viz. 50% of them carrying the gene 'T' and the other 50% with gene 't'. This shows that the two alleles T and t which are brought together in F_1, separate or segregate from each other during gamete formation.

14. These gametes fuse at random and produce tall and dwarf plants in the ratio of 3:1.
15. In the F_2 progeny, the phenotypic ratio of monohybrid cross is 3:1 and the genotypic ratio is 1:2:1 (TT:Tt:tt).

    The monohybrid cross can be represented as follows:

	Tall		Dwarf
P_1:	TT	X	tt
G_1:	(T)		(t)
F_1:	Tt	X	Tt (selfing)
	Tall		
	(T) (t)		(T) (t)

♀ \ ♂	T	t
T	TT Tall	Tt Tall
t	Tt Tall	tt Dwarf

16. Dihybrid cross is a cross between two individuals differing in two pairs of contrasting characters.
17. Mendel crossed a pea plant with round shaped seed and yellow coloured cotyledons with another pea plant with wrinkle shaped seed and green coloured cotyledons.
18. Round shape of the seed is dominant over wrinkle shape and yellow colour over green.
19. The gene governing round shape is represented by 'R' and wrinkled shape by 'r'. Similarly, yellow colour is governed by the gene 'Y' and green colour by 'y'. Therefore, the genotype of a pure round yellow individual can be represented as RR YY and wrinkled green one as rr yy.
20. The parent with round and yellow seeds produce gametes all of which carry the genes for dominant traits i.e., RY and the other parent with wrinkled and green seeds produce gametes all of which carry the genes ry. These gametes fuse at random and the offsprings produced from the seeds of such a cross were all Round and Yellow seeded with a Genotype Rr Yy. (double heterozygous). These heterozygous F_1 plants were self-fertilized.
21. Each F_1, dihybrid, during gamete formation produce four kinds of gametes in equal proportion. They are the gametes carrying RY, Ry, rY and ry genes. This indicates that at the time of gamete formation, the gene 'Y' or 'y' enters the gamete independent of 'R' or 'r' i.e., the gene 'Y' can be passed onto the gametes either with gene 'R' or 'r', and the gene 'y' can be passed with either 'R' or 'r' in equal proportion.
22. As a result of independent assortment of genes, in F_2 the four phenotypes - yellow round, yellow wrinkled, green round and green wrinkled were obtained in 9:3:3:1 ratio.
23. Therefore, the phenotypic ratio in dihybrid cross is 9:3:3;1 and the genotypic ratio is 1:2:1:2:4:2:1:2:1 (RRYY:RRYy:RRyy:RrYY:RrYy:Rryy:rrYY:rrYy:rryy).

The dihybrid cross can be represented as follows:

	Round yellow		wrinkled green
P:	RRYY	X	rryy
G_1:	(RY)		(ry)
F_1:	RrYy (heterozygous round yellow)	X	Rr Yy (selfting)
G_2:	(RY) (Ry) (rY) (ry)		(RY) (Ry) (rY) (ry)

F_2:

♀ \ ♂	RY	Ry	rY	ry
RY	RRYY Round yellow	RRYy Round yellow	RrYY Round yellow	RrYy Round yellow
Ry	RRYy Round yellow	RRyy Round green	RrYy Round yellow	Rryy Round green
rY	RrYY Round yellow	RrYy Round yellow	rrYY Wrinkled Yellow	rrYy Wrinkled Yellow
ry	RrYy Round yellow	Rryy Round green	rrYy wrinkled yellow	rryy wrinkled green

Phenotypic Ratio	**Genotypic Ratio**
Round yellow-9	RRYY-1
Round green-3	RRYy-2
Wrinkled yellow-3	RRyy-1
Wrinkled green-1	RrYY-2
	RrYy-4
	Rryy-2
	rrYY-1
	rrYy-2
	rryy-1

Mendel selfed members of the F_2 progeny and found that out of the dominant types, one-third bred true for the dominant character, whereas two-thirds segregated into dominants and recessives in the ratio of 3:1. All the recessive plants of F_2 generation when selfed bred true for the recessive character. Mendel found similar results in monohybrid crosses with all the seven pairs of contrasting characters in *Pisum sativum*. After eight years of detailed investigations on thousands of pea plants, Mendel published his results in a paper entitled "Experiments in Plant Hybridization" in the Proceedings of the Brunn Natural History Society in 1866. However, his work received no attention for 34 years until three scientists, Devries in Holland, Correns in Germany and Tschermak in Austria working independently published their findings in 1900 and confirmed Mendel's results.

The Test Cross: Not satisfied with his work, Mendel himself subjected his results to a test. In the cross between tall and dwarf pea plants, the F_1 hybrids were all phenotypically tall but their genotypes were not only *TT* but also *Tt*. Consider a heterozygous hybrid plant *Tt*. When it forms gametes, the factors *T* and *t* segregate in the gametes in a 1:1 ratio. This means that 50% of the gametes of an F_1 heterozygous hybrid carry the factor *T* and 50% the factor *t*. Mendel crossed such a hybrid plant (*Tt*) with a plant of the true breeding, dwarf variety (*tt*). All the gametes of the homozygous dwarf plant carried the recessive factor *t*. Every gamete of the recessive parent has 50% chance of combining with a gamete carrying *T* and 50% chance to combine with a *t* gamete from the heterozygous parent. This should result in 50% of progeny showing the tall phenotype and genetic constitution *Tt*, whereas 50% of the progeny should be phenotypically dwarf with genotype *tt* as explained diagrammatically below:

P:	Tall	(*Tt*) ×	Dwarf (*tt*)	
Gametes:	*T* (50%),	*t* (50%)	All	*t*
Test Cross:	Tall	:	Dwarf :	: 1 : 1
Progeny:	(*Tt*)		(*tt*)	
	50%		50%	

Indeed Mendel's results of this cross agreed with the theoretical expectations thus providing additional experimental proof of the correctness of his interpretations. Such a cross where an individual is crossed to a double recessive parent to test and verify the individual's genotype is called a *testcross* or *backcross*.

In order to determine genotypes of the F_2 progeny, Mendel allowed the F_2 plants to self-fertilise and produce a third filial or F_3 generation. He found that the homozygous F_2 tall plants could produce only tall plants on self-fertilisation. This indicated their genotype to be *TT*. Similarly the F_2 dwarf homozygotes yielded only dwarf plants on selfing; their genotype was tt. The F_2 heterozygotes on self fertilising behaved identical to the F_1 hybrids and gave rise to tall and dwarf phenotypes in the ratio 3:1. This proved that their genotype was identical to that of F_1 hybrids i.e., *Tt*.

It is noteworthy that the genotypes of the parents are written as *TT* and *tt* instead of single *T* and *t*. This is in accordance with Mendel's hypothesis that each parent has two factors for a character. There is also a cytological explanation. The somatic chromosomes of all plants and animals exist in homologous pairs, one member of each pair coming from the paternal parent, other from maternal parent. A gene is a section of the chromosomal DNA which has information necessary for determination of a specific genetic trait. Suppose a hypothetical gene *A* occupies a particular site or locus on a given chromosome. The homologous chromosome contains at the identical locus an alternative gene *a* which controls the same trait as gene *A*, but in such a way as to produce a different phenotype for the same trait. The alternative genes at the same locus *A* and *a* are also called alleles. It is an astonishing fact that though Mendel knew nothing about genes, he could predict the existence of factors, which later turned out to be genes. During the reduction division of meiosis (Metaphase I), chromosomes of a pair separate and go to the opposite poles. Consequently genes or alleles segregate from each other and pass into different gametes.

Demonstration of Genetic Segregation

Mendel's F_1 hybrids (*Tt*) were all tall plants indistinguishable phenotypically. Sometimes homozygous and heterozygous plants show phenotypic differences. There is a seedling character for green pigment in soybeans. The homozygous (*GG*) soybean plant is dark green, the heterozygous (*Gg*) plant light green. The homozygous recessive (*gg*) produces a golden lethal seedling which dies in early stages due to lack of green pigment. If the heterozygous plants are grown to maturity and self-pollinated, their progeny will again segregate as dark green, light green and lethal golden in the ratio of 1:2:1.

Differences between homozygous and heterozygous genotypes can sometimes be observed in the gametes. In rice, sorghum and maize, effect of the gene for waxy endosperm is visible in the pollen grains. Maize kernels which have waxy endosperm produce starch and stain blue with iodine; nonwaxy endosperm does not produce starch and stains red with iodine. In maize gene for waxy endosperm is located on chromosome 9. A homozygous plant with genetic constitution *Wx Wx* produces starch in endosperm and stains blue with iodine. In the heterozygous plant (*Wx wx*) the dominant gene causes starch production and the kernels stain blue with iodine. But kernels on homozygous recessive plants (*wx wx*) have no starch and stain red with iodine. If anthers of these plants are treated with iodine, the pollen grains stain in a similar way. In homozygous plants all the pollen grains stain blue. In heterozygous plants 50% of pollen grains stain blue (i.e., those containing *Wx*), whereas 50% stain red (i.e., pollen grains having *wx*). In the homozygous recessive plant, all the pollen grains stain red. If breeding tests are done by self pollinating the heterozygous F_1 plants, the F_2 progeny consists of blue staining kernels (*Wx Wx* and *Wx wx* plants) and red staining kernels (*wxwx* plants) in the ratio 3:1.

The Dihybrid Cross

Mendel made crosses between pea plants differing in two characters such as texture of seed and colour of cotyledons. Such a cross in which inheritance of two characters is considered is called a dihybrid cross.

First of all Mendel crossed a pea plant that was breeding true for round seeds with a plant that bred true for wrinkled seeds. The F_1 indicated that roundness was dominant over wrinkled texture of seed coat. Similarly, by another cross he could determine that yellow colour of cotyledons was dominant over green. He now used as male parent a plant which bred true for both round and yellow characters and crossed it with a female parent that bred true for wrinkled green. As expected from the results of his single crosses, the F_1 was round yellow. When he selfed the F_1 hybrids, the F_2 progeny showed all the parental characters in different combinations with each other. Thus plants with round yellow seeds, round green seeds, wrinkled yellow seeds and wrinkled green seeds all appeared in the ratio 9:3:3:1. Reciprocal cross in which the female parent was round yellow and male parent wrinkled green gave the same results.

Mendel applied the principles of a monohybrid cross and argued that in the dihybrid cross the true breeding yellow parent must be homozygous *RRYY*, and the wrinkled green parent *rryy*. Since each character is determined by two factors, in a dihybrid cross there must be four factors present in each parent. Likewise the F_1 hybrid must be *RrYy*. But the question remained as to how did the four different combinations of parental phenotypes appear in the progeny? Mendel argued that the pair of factors for roundness must be behaving independently of the pair of factors for yellow colour of seeds. In other words, one factor for a character must be passing independently of a factor for another character. Thus in the F_1 hybrids, *R* and *r* pass into different gametes. Now the probability of an *R* gamete formed is one-half, and of *r* gamete and one-half. Similar probabilities exist for *Y* and *y* gametes. It follows that the probability that *R* and *Y* should go to the same gamete is one-fourth, as also of *R* and *y*, *r* and *Y*, and *r* and *y*. Therefore, gametes containing factors *RY*, *Ry*, *rY* and *ry* should form in equal proportions.

The F_1 hybrid producing the four types of gametes mentioned above was selfed. The results expected in the F_2 progeny can be predicted by making a checkerboard or a Punnett Square. Gametes produced by one parent are plotted on top of the checkerboard, and gametes of the other parent on the side. The sixteen squared of the checkerboard are filled up by making various possible combinations of male and female gametes during fertilisation. The phenotypes read out from the checkerboard indicate a 9:3:3:1 ratio exactly as observed by Mendel.

P: Round Yellow × Wrinkled Green

RRYY *rryy*

Gametes: *RY* *ry*

F_1: Round Yellow × Self (Round Yellow)

RrYy *RrYy*

Gametes: *RY, Ry, rY, ry* *RY, Ry, rY, ry*

	RY	Ry	rY	ry
RY	RRYY	RRYy	RrYY	RrYy
Ry	RRYy	RRyy	RrYy	Rryy
rY	RrYY	RrYy	rrYY	rrYy
ry	RrYy	Rryy	rrYy	rryy

As in the case of the monohybrid cross, Mendel verified his results by performing the test cross. He crossed the F_1 hybrid heterozygous for both characters with a double recessive parent (*rryy*) which should produce only one type of gamete *ry*. The uniformity in the gametes of the recessive parent determines the differences in the types of gametes produced by the heterozygous parent. Now the hybrid *RrYy* produces gametes carrying *RY, Ry, rY* and *ry* with equal frequency. It follows that during fertilisation if *all* these four types of gametes unite with *ry* gamete of the recessive parent, the resulting progeny should show all the four combinations of characters also in equal proportions. Indeed, Mendel observed the testcross progeny to consist of Round Yellow, Round Green, Wrinkled Yellow and Wrinkled Green plants in the ratio 1:1:1:1.

F_1: Round Yellow × Wrinkled Green

RrYy *rryy*

Gametes: *RY, Ry, rY, ry* *ry*

		RY	Ry	rY	ry
F_2:	ry	RrYy	Rryy	rrYy	rryy

From the results of his dihybrid crosses, Mendel realised the following facts. At the time of gamete formation the segregation of alleles *R* and *r* into separate gametes occurs independently of the segregation of alleles *Y* and *y*. That is why the resulting gametes contain all possible combinations of these alleles, i.e., *RY, Ry, rY, ry*. In this way Mendel proved that when two characters are considered in a cross, there is independent assortment of genes for each characters, and this became the *Law of Independent Assortment.*

Trihybrid and Multihybrid Crosses

Mendel extended his observations to trihybrid crosses involving three pairs of contrasting characters. The characters he considered were: seed shape – smooth (*S*) vs. wrinkled (s); colour of cotyledons – yellow (*Y*) vs. green (*y*); and flower colour – violet (V) vs. white (*v*).

P:	Smooth Yellow Violet	×	Wrinkled Green White
	SSYYVV		*ssyyvv*
Gametes:	*SYV*		*syv*

F_1: Smooth Yellow Violet × Self

SsYyVv

Gametes: SYV, SyV, SYv, Syv

sYV, syV, sYv, syv

F_2: 27 Smooth Yellow Violet: 9 Smooth Yellow White:
9 Smooth Green Violet: 9 Wrinkled Yellow Violet:
3 Smooth Green White: 3 Wrinkled Yellow White:
3 Wrinkled Green Violet: 1 Wrinkled Green White.

Incomplete Dominance:

A monohybrid cross between a red flowered snapdragon (Antirrhinum majus) and a white flowered variety does not produce red or white flowered plants in F_1 as expected from mendelism. Instead the flowers are pink, i.e., intermediate between the two parents. This is because neither red flower colour nor white is dominant, but each allele has its influence in colour development and the hybrid appears pink. If the F_1 pink flowers are self-pollinated, the F_2 progeny shows red, pink and white flowered plants in the proportion 1:2:1. It may be recalled that this is the same genotypic ratio that Mendel obtained in garden peas. The difference is that in the present case the heterozygous progeny is distinct in appearance from the homozygotes. The term intermediate inheritance is also given to crosses where F_1 hybrids show incomplete dominance or partial dominance with no phenotypic resemblance to either parents.

EXERCISE:

1. Write in brief Mendel's works and explain with example the test cross.
2. Write in detail with example, incomplete dominance.
3. Write short notes on:

(i) Monohybrid cross.
(ii) Dihybrid cross.
(iii) Trihybrid and Multihybrid crosses.
(iv) Phenotype and genotype.
(v) Significance of the test cross.

4. A cross between yellow round and green wrinkled seeded pea plants gave the following 4 types of progeny in equal proportions: yellow round, yellow wrinkled, green round and green wrinkled. What are the genotypes of the parents.

10

Interaction of Genes

INTERACTION OF GENES:

According to Mendel, the two alleles governing a particular character are generally related as dominant and recessive, so that in a heterozygous condition, the dominant character is expressed and the recessive one is suppressed and hence the heterozygous individual resembles the dominant parents.

But several cases were recorded where the heterozygous individuals were not related to either of the parents, but exhibited an intermediate character of the two parents. In such individuals, the dominant allele in heterozygous condition has reduced expression, so that each of them expresses itself partially. This process where one allele is completely dominant over the other so that in a heterozygous both alleles are partially expressed resulting in intermediate phenotype is called as Incomplete dominance. The intermediate phenotype in a heterozygote is due to the interaction between the two alleles.

EXAMPLES:

(a) In Mirabalis jalapa (4' O clock plant), when plants with red flowers (RR) are crossed with plants having white flowers (rr), the F1 hybrid plants bear Pink flowers (Rr). When these F_1 hybrids are selfed, they produce Red (RR), Pink (Rr) and White (rr) flowered plants in the ratio of 1:2:1. The cross can be represented as follows:

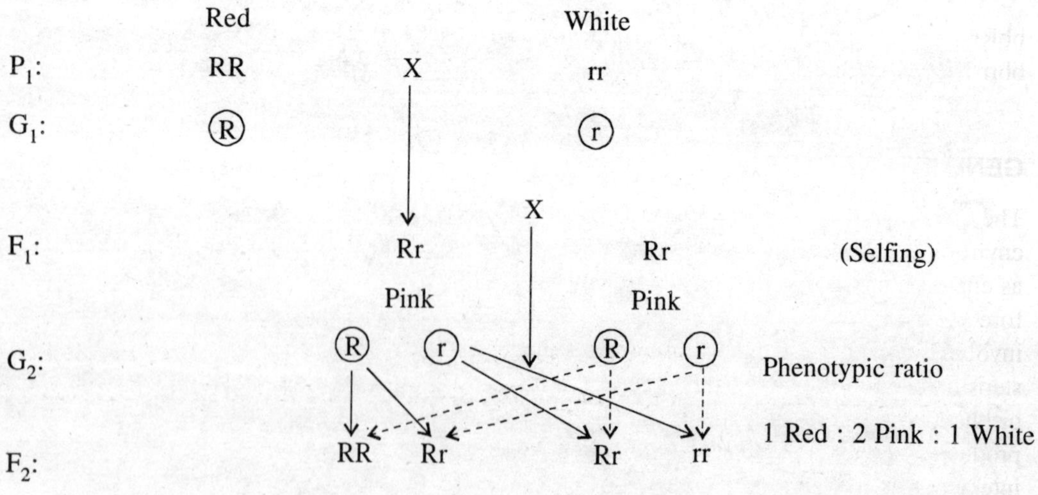

(b) In Antirrhinum majos (Snapdragon), the gene governing Broad leaves (B) is incompletely dominant over its allele b governing short/narrow leaves and the gene governing Red colour of flower (R) is incompletely dominant over its allele for white flower (r), so that in heterozygous condition, Bb produces intermediate leaves and Rr produces Pink flowers. If a dihybrid cross is made between plants with Broad leaves and Red flowers and with narrow leaves and white flowers, the F_1 produced is a dihybrid with intermediate leaves and Pink flowers. When these F_1 plants were selfed, nine phenotypes corresponding to nine genotypes were obtained in 1:2:1:2:4:2:1:2:1. The cross can be represented as follows:

	Broad Red		Narrow White
P_1:	BBRR	X	bbrr
G_1:	(BR)	↓	(br)

F_1: BbRr (Intermediate, Pink) X BbRr ↓

F_2

♀ \ ♂	BR	Br	bR	br
BR	BBRR Broad Red	BBRr Broad Pink	BbRR Inter, Red	BbRr Inter, Pink
Br	BBRr Broad Pink	BBrr Broad White	BbRr Inter, Pink	Bbrr Inter, White
bR	BbRR Inter, Red	BbRr Inter, Pink	bbRR Narrow Red	bbRr Narrow Pink
br	BbRr Intermediate Pink	Bbrr Inter, White	bbRr Narrow Pink	bbrr Narrow White

Phenotypic Ratio:

BBRR-Broad Red-1

BBRr-Broad Pink-2

BBrr-Broad White-1

BbRR-Intermediate Red-2

BbRr-Inter, Pink-4

Bbrr-Inter, White-2

BbRR-Narrow Red-1

bbRr-Narrow Pink-2

bbrr-Narrow White-1

GENE INTERACTION:

The phenotype is a result of gene products brought to expression in a given environment. The environment includes not only external factors such as temperature etc., but also internal factors such as enzymes and hormones. Genes specify the structure of Protein. Most enzymes are proteins. Therefore each enzyme is coded by a gene. Several genes are usually required to specify the enzymes involved in even the simplest pathways. Whenever two or more genes specify enzymes that catalyse steps in a common pathway, then genetic interaction can occur. Therefore, when a particular character or phenotypic expression is governed by more than one pair of genes, the non-allelic genes interact to produce a new phenotype. This interaction between non-allelic genes is called Gene Interaction Gene interaction is of two types.

(A) Epistatic Gene Interaction

(B) Non-Epistatic Gene Interaction.

EPISTATIC GENE INTERACTION OR EPISTASIS:

Epistasis is the interaction between non-allelic genes in which one gene inhibits or suppresses the expression of the other gene. The gene that suppresses or masks the expression of the other gene is known as epistatic gene and the locus where it is present is called a epistatic locus. The gene that is suppressed by the epistatic gene is called as hypostatic gene and the locus as hypostatic locus. Epistasis involves intergenic suppression. When epistasis is operative between two geneloci, the number of phenotypes appearing in the offspring from dihybrid parents will be less than 4.

There are six types of epistatic gene interactions. They are:

1. Dominant Epistasis
2. Recessive Epistasis
3. Duplicate Dominant Epistasis
4. Duplicate Recessive Epistasis
5. Duplicate Genes with cumulative effect
6. Dominant and Recessive interaction.

Duplicate Recessive Interaction: Complementary Genes:

Two pairs of non-allelic dominant genes interact to produce only one phenotypic trait, but neither of them if present alone in dominant condition can produce the phenotypic trait. Absence of even one of the two genes in dominant condition produces recessive phenotype i.e., different pairs of homozygous recessive genes suppress the expression of phenotypic trait. Hence, 9:3:3:1 ratio is modified into 9:7 ratio.

Example: (a) Flower color in sweet pea

In Lathyrus odaratus (sweet pea), the purple color of flowers is dependent on two nonallelic complementary genes C and P. Gene C Produces an enzyme that catalyzes the formation of colorless chromogen for the formation of anthocyanin pigment. Gene P controls the production of an enzyme which catalyze the transformation of this chromogen into anthocyanin.

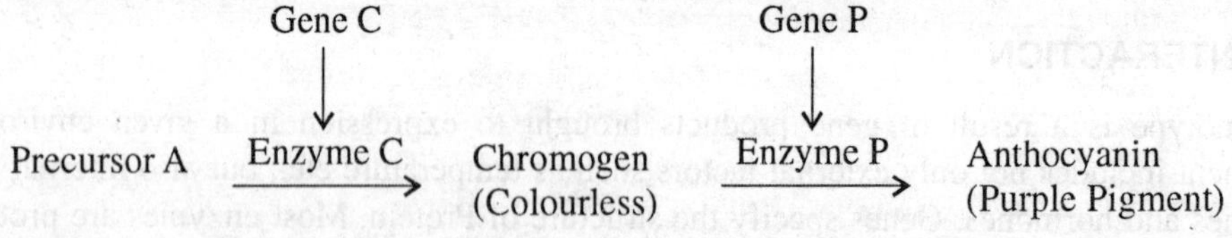

Therefore, a plant produces purple flowers if both dominant genes C and P are present. If any one of the two genes is in recessive condition. White flowers are produced as anthocyanin pigment is not formed.

Bateson artificially crossed two white strains of Lathyrus, the progeny had purple flowers. These purple individuals when self pollinated produce purple and white in 9:7 ratio.

White   White

CCpp   X   ττPP

(Cp)   (τP)

CτPp
Purple   X   CτPp   (Selfing)

♀ \ ♂	CP	Cp	τP	τp
CP	CCPP Purple	CCPp Purple	CτPP Purple	CτPp Purple
CP	CCPp Purple	CCpp White	CτPp Purple	Cτpp White
τP	CτPP Purple	CτPp Purple	ττPP White	ττPp White
τp	CτPp Purple	Cτpp White	ττPp White	ττpp White

Ratio   C - P → 9   Purple

C - pp → 3

ττP - → 3   7 white

ττpp → 1

Dominant and Recessive Interaction: Epistasis

In this type of interaction, the dominant genotype at one locus (A-) and the recessive genotype at other locus (bb) produce the same phenotypic effect. Thus A- B-, A - bb, aa bb produce one phenotype and aaB - produce another in the ratio 13:3.

Example: Plumage color Fowls: In poultry, there are 2 kinds of white breeds. They are the white leghorn breed and white plymouth rock breed. The coat color is controlled by the action of two gene pairs I and C. The white color of leg horn breed is due to the dominant gene 'I', while the white color of plymouth rock breed is due to its recessive genotype iiττ.

When a leg horn breed (IICC) of fowl is crossed to plymouth rock breed (iiττ), in F_1 all are white and when selfed produce white and colored fowls in 13:3 ratio.

	White leg horn		White plymouth rock
P₁:	IICC	X	iiττ
G₁:	(IC)	↓	(iτ)
F₁:		IiCτ	X IiCτ (Selfing)

♀ \ ♂	IC	Iτ	iC	iτ
IC	IICC White	IICτ White	IiCC White	IiCτ White
Iτ	IICτ White	IIττ White	IiCτ White	Iiττ White
iC	IiCC White	IiCτ White	iiCC Colored	iiCτ Colored
iτ	IiCτ White	Iiττ White	iiCτ Colored	iiττ White

Ratio I - C → 9
I - ττ → 3
iiττ → 1 } 13 White
iic - → 3 Colored

Table: Summary of Epistatic Ratios:

Genotype	*A-B-*	*A-bb*	*aaB-*	*aabb*
Classical Dihybrid Ratio	9	3	3	1
Dominant Epistatis	12		3	1
Recessive Epistasis	9	3	4	
Duplicate Dominant interaction		15		1
Duplicate recessive interaction	9		7	
Duplicate genes with cumulative effect	9		6	1
Dominant and Recessive interaction	13		3	

NON EPISTATIC GENE INTERACTION: SUPPLEMENTARY FACTORS

Gene interaction may also occur without epistasis. Interaction of two non-allelic genes influencing the same character can give rise to entirely new phenotypes that neither of the genes by itself could produce. This was discovered by Bateson and Punnet in Fowls.

EXAMPLE:

Comb pattern in Fowls: The development of Comb is controlled by two independently assorting non-allelic gene pairs R and P. Gene R gives rise to Rose comb and Gene P produces pea comb. Genes R and P for rose and pea combs together produce a new phenotype i.e., walnut comb while single comb is produced when both R and P genes are in recessive condition.

Wyandotte variety of domestic chicken possesses rose comb, while Brahmas have pea comb. When these two varieties are crossed, F_1 chicken developed Walnut comb, phenotype not expressed in either parent. When F_1 are mated among themselves, in F_2 Walnut, Rose, Pea and Single combed chicken are obtained in 9:3:3:1 ratio. The cross can be represented as follows:

Rose comb Pea comb

RRpp X rrPP

(Rp) ↓ (rP)

RrPp X RrPp (Selfing)

Walnut ↓

♀ \ ♂	RP	Rp	rP	rp
RP	RRPP Walnut	RRPp Walnut	RrPP Walnut	RrPp Walnut
Rp	RRPp Walnut	RRpp Rose	RrPp Walnut	Rrpp Rose
rP	RrPP Walnut	RrPp Walnut	rrPP Pea	rrPp Pea
rp	RrPp Walnut	Rrpp Rose	rrPp Pea	rrpp Single

Ratio R_ P- → 9 Walnut

R-pp → 3 Rose

rrP- → 3 Pea

rrpp → 1 Single

MULTIPLE ALLELES:

Mendel identified just two unlike members in each gene pair. These are called alleles. These alleles or allelomorphs occupy the same position or locus on the homologous Chromosome. It is supposed that one allele arose as mutant from one of the identical alleles. If one of the two dissimilar alleles mutates, a third variation could be noticed. Yet another mutation may produce a fourth variation and so on. These alleles which occupy the same locus on the homologous chromosomes are called Multiple alleles. Any individual contains just two alleles. Hence multiple alleles can be defined as "a set of three or more alleles which arise as a result of mutation of a normal gene and which occupy the same locus on homologous chromosomes."

Characters of Multiple Alleles:

1. Multiple alleles occupy the same locus within the homologous chromosomes only one member of the series is present in a given chromosome.
2. Since only two chromosomes of each type are present in diploid cell, only two alleles of the multiple allelic series are found in a cell and also in a given individual.
3. In a gamete only one allele of multiple allelic series will be present as each gamete contain only one chromosome of each type.
4. There is no crossing over within multiple allelic series.
5. A set of multiple alleles control the same primary character but each of them is characterised by different manifestation.
6. The normal allele is dominant over all other mutant alleles. The intermediate members of the series may be related as dominant and recessive or they may even exhibit codominance.

EXAMPLE: MULTIPLE ALLELES IN BLOODS GROUPS OF MAN:

Landsteiner discovered different blood groups in man. He noted that a smooth combination always resulted when one's cells were recombined with one's own plasma, but when plasma from one person was mixed with cells from another person the mixture would sometimes be smooth but in other cases would result in clumping of the cells. This led to the discovery of blood groups. There are four different types of blood groups in man. They are A, B, AB and O.

The difference in blood groups is according to the presence or absence of antigens and antibodies. Antigens are proteinaceous substances which are capable of stimulating the production of specific antibodies. Antibodies are substances produced by animals in response to contact with foreign antigens and react specifically to particular antigens.

Landsteiner found that the RBC of man contain two distinct antigens and the letters A and B were chosen to represent them. Persons with type A blood possess the A antigen and b antibodies, type B person have B antigen and a' antibodies, type AB person have both A and B antigens and no antibodies and person with type 'O' blood possess neither of these antigens but have both 'a' and 'b' antibodies. These blood groups are important in blood transfusion because no blood should be given to a person whose plasma contains antibodies that will clump the red cells as fast as they enter his body.

Inheritance of blood types: Antigens are produced by an autosomal gene. When a person carries the gene for A antigen his blood will contain A antigen and if he carries the gene for B antigen his blood will contain B antigen. When a person carries the genes for A antigen and B antigen, his blood will contain both antigens. Hence neither gene is dt. over other. Each of the blood group alleles seems to act independently and is not suppressed by the presence or absence of any of the other allele. The letter 'I' (isohaemogglutinogen) is used as the basic symbol for the genes at this locus, with a second letter as an exponent to indicate which variation of the allele is represented. For Example. I^A represent the gene which produces 'A' antigen and I^B represents the gene which produces B antigen. Persons heterozygous for these two genes will have both antigens in their blood i.e., I^AI^B and it is classified as type AB. The type of 'O' individuals will have homozygous recessive genes (i.e.) I^OI^O or ii.

The various genotype of different blood groups can be shown by the following table

Blood Group	Genotype	Antigens	Antibodies
A	I^AI^A or I^AI^O	A	b
B	I^BI^B or I^BI^O	B	a
AB	I^AI^B	A and B	Nil
O	I^OI^O	No antigen	a and b

Multiple Factors: Skin Colour in Human Beings:

Multiple factors is referred to involvement of many genes in the inheritance of a particular character. But according to Fraser (1976), Multiple factors or polygenes should refer to conditions determined by a large number of genes each with a small effect acting additively. Multiple factors – inheritance is determined by a combination of genetic and environmental factors. However, it is difficult to ascertain whether or not environmental factors are involved, or whether all the genes controlling a trait have little and additive effects.

Skin colour inheritance in human beings: Variations occur in human population with reference to skin colour; this prevails in much greater degree between different populations. Human skin colours have been analysed by many workers and human races have been established as white, black, yellow and red on the basis of skin colour. The Africans and the Europeans who show both extremes of black and white skin respectively have been extensively studied. The children and grand-children from marriages between negroes and Europeans give rise to hybrid groups showing quantitative variations in skin pigment. In 1913 Davenport and Danielson and later Gates, Curtstern and others have conducted investigations of this nature.

Among these studies, with marriages between whites and negroes, the F_2 generation was found to be intermediate in skin colour to both parents and was called a mulatto. Davenport revealed that crosses between two mulattos produced shades of colour falling in five distinct classes. Davenport therefore considered two gene pairs to control skin colour inheritance in human beings. Later on Gates assigned 3 genes, and Stern 4 to 6 genes for this trait. Harrison and Owen in 1964 have estimated the number of genes, using modern techniques of reflectance spectrophotometry, for skin colour to be between 3 and 4.

A child with white European skin would be expected in one birth out of every 256, and the same probability is true for negro skin colour, presuming that 4 genes determine skin colour.

EXERCISE:

1. A rose combed chicken is crossed with a walnut. The progeny consists of 3 walnut, 3 rose, one pea and one single. What are the phenotypes of the parents?
2. Four hens with walnut combs are each mated with a single combed cock. One mating produces only walnut comb; the second produces both walnut and peacomb; the third produces rose and walnuts and the fourth produces walnut, rose, pea and single combs. What are the genotypes of the four walnut hens used as parents?

3. In sweet peas the gene 'C' and 'P', when present together, produce purple flowers. But when either 'C' or 'P' is present alone it produces white flowers. What flower colour could the progeny of the following crosses have
   (a) Ccpp × ccPP
   (b) ccPp × CcPp
   (c) CCPp × ccPP
4. Justify that human blood groups are governed by multiple alleles.
5. Justify that human skin colour is governed by multiple factors.
6. Write short notes on
   (a) Epistosis
   (b) Multiple alleles
   (c) Multiple factors
   (d) Gene interaction.

11

Sex Determination in Plants and Animals

1. The sexually reproducing organisms are classified into two types. They are monoecious and Dioecious. In monoecious organisms (bisexual) both male and female gametes are produced by a single individual while in dioecious or unisexual organisms, male and female gametes are produced by different individuals. The individuals which produce only male gametes are called as male individual and those which produce female gametes arc called as female individual. The phenomenon of molecular, morphological, physiological differentiation between male and female sexes is called sexual dimorphism.
2. In sexually dimorphic organisms, the sexual diversity also occurs at the level of chromosomes. In these organisms, the chromosomes can be grouped into two categories. The are
   (a) **Autosomes:** The chromosomes which have no relation with the sex and contain genes which determine the somatic characters of the individuals are known as **autosomes**. The number of autosomes may vary in the individuals of different species.
   (b) **Allosomes or Sex chromosomes:** The chromosomes which are responsible for the determination of sex are known as **allosomes**. Only one pair of sex chromosomes are present in an individual X and Y-chromosomes are called as sex chromosomes.

Chromosomal basis of Sex Determination:

3. The chromosomal differences in the two sexes play a role in determination of sex in most of the dioecious organisms. The sex chromosomal determination can be differentiated into two types. (a) Heterogametic males or male digamety (b) Heterogametic females or female digamety.

Heterogametic males:

4. In this type the males are heterogametic and produce two types of gametes i.e., 50% of the gametes carry 'X'-chromosome while the other 50% of the gametes lack 'X' chromosome.
5. The females have two 'X'-chromosomes and hence are homogametic and produce gametes all of which carry one 'X'-chromosome.
6. Heterogametic males are of two types (i) ♀ XX - XO♂ type (ii) ♀ XX - XY♂ type.
7. XX-XO type of sex determination is observed in some insects belonging to orders Hemiptera (Eg., Beg bugs) and orthoptera (Eg., Grasshoppers). The females possess two X-chromosomes and

males have only one 'X'-chromosome. There is no 'Y'-chromosome in males and hence the sex chromosomal composition in males is represented as 'XO'. Fifty percent of sperms receive 'X' chromosomes while the other fifty percent are devoid of any sex chromosome and contain only haploid number of autosomes. If the sperm carrying 'X'-chromosome fertilizes an egg, it will be a female while a sperm lacking 'X'-chromosome fertilizes an egg, it will be a male.

8. XX-XY type of sex determination is the most common one and is observed in Drosophila, Man, other mammals, Melandrium (plant) etc. In this, the females are homogametic as they contain two 'X' chromosomes while males contain one 'X' chromosome and one 'Y' chromosome. 'Y' chromosome differs from 'X'-chromosome in its morphology.
9. The 'Y' chromosome may be larger than 'X'-chromosome as in Drosophila or smaller than 'X' chromosome as in Man. 'X' and 'Y' chromosomes possess homologous and differential regions i.e., partly homologous and partly non-homologous. The differential region influences the sex determination. The male having two heteromorphic sex chromosomes produce two kinds of sperms i.e., 50% with 'X'-chromosomes and other 50% with Y. The sex of the embryo depends on the kind of sperm. An egg fertilized by an 'X' bearing sperm produces a female and if fertilized by a Y-bearing sperm produces a male.

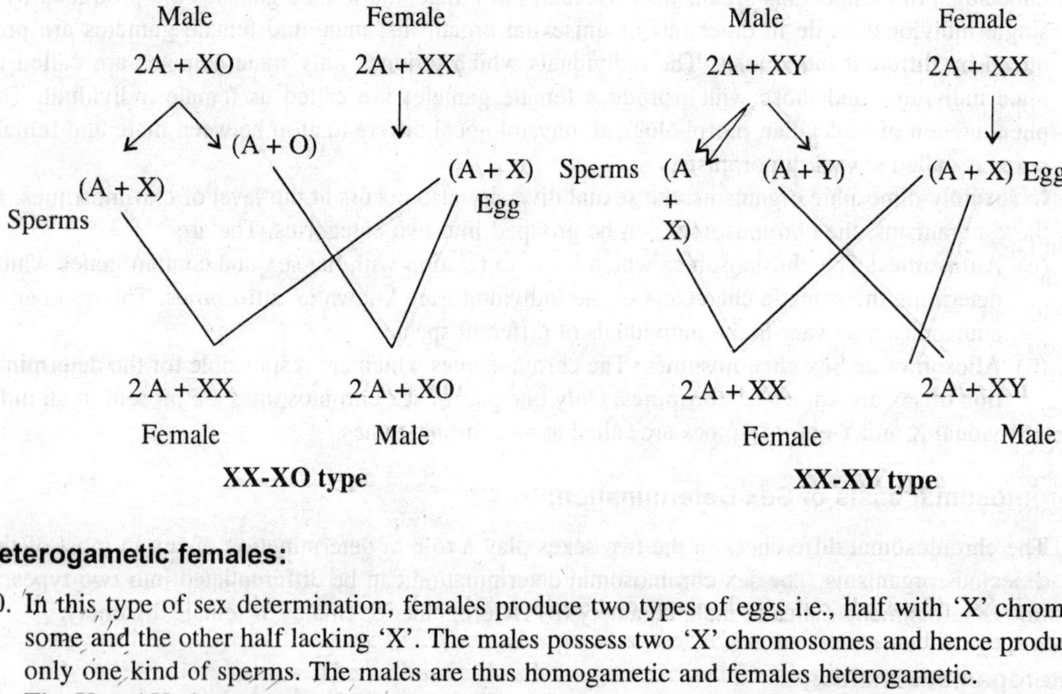

XX-XO type **XX-XY type**

Heterogametic females:

10. In this type of sex determination, females produce two types of eggs i.e., half with 'X' chromosome and the other half lacking 'X'. The males possess two 'X' chromosomes and hence produce only one kind of sperms. The males are thus homogametic and females heterogametic.
11. The X and Y chromosomes in heterogametic females are often represented as Z and W chromosomes respectively.
12. Heterogametic females are of two types (a) ZO-ZZ type and (b) ZW-ZZ type.
13. ZO-ZZ type or XO-XX type of sex determination is observed in butterflies, certain months and chicken. The females produce two kinds of eggs, 50% of them carrying Z or X-chromosome and other 50% lacking it. The sex of the offspring depends on the kind of egg which undergoes fertilization. If a Z bearing egg fertilizes a Z bearing sperm it will be a male while if an egg lacking Z fertilizes a Z bearing sperm it will be a female.

14. ZW-ZZ type or XY-XX type of sex determination is observed in birds, fishes, reptiles and certain insects. The sex chromosomal complement of females is represented as ZW and males as ZZ. Hence females produce two kinds of eggs – 50% with Z and 50% of with W chromosome, while males produce only one kind of sperms, all of them carrying a Z chromosome. The sex of the offspring depends on the kind of egg involved in fertilization with Z bearing sperm, Z bearing egg producing males and W bearing eggs producing females.

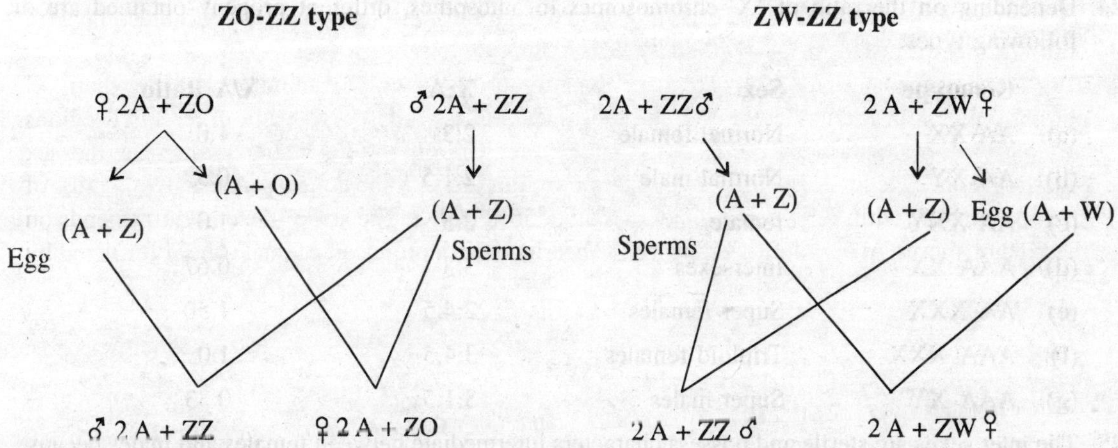

Genic Balance Theory of Sex Determination in Drosophila:

15. In Drosophila, apart from the chromosomal basis of sex determination, a more complex process in the determination of sex is involved which is called as **genic balance mechanism.**
16. Morgan's work on Drosophila indicated that many sex linked genes are present in 'X' chromosome and Y is an empty chromosome without any genes.
17. Later, C.B. Bridges found a male Drosophila without any Y-chromosome i.e., XO. These males were sterile. He also found some females with two 'X'-chromosomes and a Y-chromosome i.e. XXY. These females were normal and fertile. He concluded from these observations that the Y chromosome in Drosophila is sexually neutral as it is not involved in Sex determination but controls male fertility.
18. The investigations on Drosophila by Bridges showed that female determining genes were located on the 'X'-chromosome and male determining genes were on the autosomes instead of Y chromosome.
19. No specific loci on the autosomes were identified determining maleness at that time, but recent evidences demonstrate that male determining genes were located on all the three autosomal chromosomes.
20. Therefore, to explain the mechanics of sex determination in Drosophila, the genetic balance theory of sex determination was devised.
21. Bridges experimentally produced various combinations of X-chromosomes and autosomes in Drosophila and deduced from comparisons that one X-chromosome and two sets of autosomes produced a normal diploid male while two X-chromosomes and two sets of autosomes produced a normal diploid female.
22. In Drosophila, sex is determined polygenically. The sex of the fly depends upon the ratio of X-chromosomes to autosomes. If each haploid set of autosomes carries factors with a male determining

value equal to one them each X-chromosome carries factors with a female determining value equal to one and a half.

23. Therefore, in a normal male (2A + XY) the male and female determinants are in the ratio $2:1\frac{1}{2}$(2A:1X) and therefore the genic balance is in the favour of maleness. A normal female (2A + XX) has a male and female determinant ratio of 2:3 (2A:2X) and therefore the genic balance is in the favour of femaleness.
24. Depending on the ratio of 'X'-chromosomes to autosomes, different progeny obtained are of following types.

	Genotype	Sex	X:A	X/A Ratio
(a)	AA XX	Normal female	2:3	1.0
(b)	AA XY	Normal male	2:1.5	0.5
(c)	AA XXY	female	2:3	1.0
(d)	AAA XX	Intersexes	3:3	0.67
(e)	AA XXX	Super females	2:4.5	1.50
(f)	AAA XXX	Triploid females	3:4.5	1.0
(g)	AAA XY	Super males	3:1.5	0.33

25. The inter sexes are sterile and possess characters intermediate between females and males because the genic balance ratio is 3:3.
26. The sex of the fly can also be observed by X/A ratio. For normal males X/Y ratio is 0.5 and 1.0 for normal females, while intersexes have X/Y ratio is between 0.5 to 1.0, super females have more than 1.0 and less than 0.5 for super males.

Environmental factors and Sex determination:

27. In some lower animals, sex determination is nongenetic and depends on factors in the external environment. Eg: Bonellia viridis, Chrysema picta, Agama agama.
28. Males and females have similar genotypes, but stimul; from environmental sources initiate development towards one sex or the other. The genetic potential for both maleness and femaleness is present in every zygote but some specific factor in the environment triggers the expression of either ♂ or ♀ phenotypes.
29. Males of the marine worm Bonellia are small and degenerate and live within the reproductive tract of the larger female. All organs of male worm are degenerate except those of the reproductive system.
30. F. Baltzer found that any young worm reared from a single isolated egg become a female. If hatched worms are released into water containing mature females, some young worms were attracted to females and became attached to the female proboscis and these were transformed into males and eventually migrated to the female reproductive tract where they become parasitic.
31. Therefore, Genetic determiners for both sexes are present in all young worms. Extracts made from the female proboscis influence young worms towards maleness.
32. In Chrysema and Agama, the temperature at which the fertilized eggs are incubated prior to hatching plays a major role in the determination of sex of the offspring.
33. In the turtle chrysema picta, high egg incubation temperature result in the production of female progeny while in lizard Agama agama, high incubation temperatures result in male offsprings.

Sex determination in plants:

34. In higher plants, the plants are classified as males, females and hermaphrodites on the basis of whether the flower possesses only anthers or only ovary or both. It is believed that the primitive type of floral condition in these plants was hermaphroditism where flowers are bisexual and advanced condition in unisexual.
35. Male and female flowers may be present on the same plant (Monoecious condition) or present on different plants (Dioecious condition) or present on same plant along with bisexual flowers.
36. If bisexual flowers and male flowers are present on same plant, the condition is called as Andromonoecious and if bisexual flowers and female flowers are present on same plant, it is called as Gynomonoecious condition.
37. Bisexual flowering plants possess potentiality for both sexes and develop both types of reproductive organs.
38. Allen reported the chromosomal basis of sex determination in certain plant species where XX-XY or XX-XO type of sex determination is observed.
39. **Examples:** (a) XX♀- XY♂ type: Melandrium album, M. rubrum, Humulu slupulus, Rumex angiocarpus, Similax cannabis etc.

    (b) XX♀- XO♂ Vallisneria spiralis and Dioscorea sinuata

    (c) XY♀- XX♂ species of fragaria
40. In certain other plants like spinacea oleracea, Ribesalpinum, Vitis cinerea, carica papaya, Asparagus officinalis, viscum fischeri etc., there are no sex chromosomes but certain gene loci control the sex determination.

SEX CHROMATIN BODIES AND DOSAGE COMPENSATION:

1. In 1949 Barr and Bertram made the important discovery that in interphase nucleus of female mammals there is a small, darkly stained chromatin body (first observed in the nerve cells of female cats) which was absent in males. This was called as sex chromatin body or Barr body after the name of its discoverer or more recently as X-Chromatin.
2. The X-chromatin can be found as a small body in different positions within the nucleus. For example, in nerve cells it may be near the nucleolus or in the nucleoplasm or near the nuclear envelope. In the cells of buccal mucosa it is generally attached to the nuclear envelope and in leukocytes it may appear as a small rod called the drumstick.
3. The frequency with which sex chromation can be detected in the female varies from tissue to tissue. In nervous tissue the frequency may be 85%. Where as in whole mounts of amniotic or chorionic epithelium it may be as high as 96%. In buccal smears the frequency may vary between 20-50% in normal female.
4. In females where the barr body is present in the nuclei of their body cells are referred as sex chromatin positive and in males where barr body is absent are referred as Sex chromatin negative.
5. The barr body is related to the number of X-chromosomes.
6. The sex of the human embryos can be distinguished at early stages of development by observing the barr bodies i.e., +ve → female and if −ve → male.
7. Sex chromatin bodies are also useful in diagnosing various kinds of sex chromosomal abnormalities in humans.
8. The individuals with two or more X-chromosomes have one less sex chromatin body than the number of X-chromosomes present.

9. Cells of abnormal females with only one X-chromosome i.e., Turner's Syndrome (XO) have no sex chromatin bodies, while cells of males with two X and one Y-chromosomes (Kline felter's Syndrome-XXY) have one sex chromatin body and cells of abnormal females with three X-chromosomes have two sex chromatin bodies in their cell nuclei.
10. Sex chromatin body is formed as a result of inactivation of one of the X-chromosome in female.
11. In mammals, the females have two X-chromosomes and the males have only one X-chromosome in the diploid cell. This means that the genes of X-chromosome are present in double dose in females where as in single dose in males. This difference of dosage in the two sexes may lead to genic imbalance in one of them. But geneticists have observed that in females homozygous for genes on the X-chromosomes do not express a trait more markedly than the hemizygous males. Thus, there must be a mechanism through which the effective dosage of genes of the two sexes is made equal or nearly so. This process is called as **Dosage compensation.**
12. This compensation in dosage of genes is achieved either by hypo production due to inactivation of one X-chromosome in females as in mammals or due to hyper production due to hyperactivity of X-Chromosome in males as in Drosophila.
13. The term Dosage compensation was first coined by Muller in 1932 while the hypothesis explaining dosage compensation was named after Mary. F. Lyon as Lyon's hypothesis, who first formulated it from cytological observations and genetic studies on coat colour in mice.
14. Female mice heterozygous for certain coat colour genes show a mottled effect unlike the homozygotes and very distinct from a uniform intermediate coat colour. Normal male mice never had the mottled effect.
15. The Lyon hypothesis was based on the observation that the number of sex chromatin bodies in interphase cells of adult females is one less than the number of X-chromosomes observed in metaphase preparations.
16. Therefore, the chromatin body is a heterochromatinised X-chromosome with facultative type of heterochromatin unlike constitutive heterochromatin found in other chromosomes.
17. Only one X-chromosome is required for normal metabolism in cells of females. Hence any additional X-chromosomes are condensed into facultative heterochromatin and becomes genetically inactive.
18. The Chromosome that becomes inactive is either maternal or paternal origin and is a matter of chance, but once an X has become inactivated, all cells arising from that cell will keep the same inactive X-chromosome. Thus mammalian X-chromosome has the capability of being hetero chromatic in some cells and euchromatic in others.
19. The fact that X inactivation occurs at random has been demonstrated in human diseases linked to the X-chromosome. For example, the Lesch-Nyhan syndrome, in which a deficiency of one enzyme HGPRT (Hypoxanthine Guanine Phospho Ribosyl Transferase) of Purine metabolism produces in mental retardation and increased uricacid levels, results from a recessive mutation in the X-chromosome. The fibroblasts of these patients when cultured in Vitro, two types of cell clones are obtained. Half of the clones contain the enzyme while the other half in which the X carrying the normal gene is condensed lack the enzyme.
20. There is no general rule for the time at which inactivation of X-chromosome occurs leading to formation of facultative heterochromatin. In mammals, inactivation occurs early in embryogenesis i.e., by 16th day of gestation in man.
21. All the genes on the condensed X-chromosome are inactive with the exception of those present on short arm showing that the inactivation of X-chromosome is not complete and that only a part of X-chromosome condenses into a bar corpuscle.

22. Once the inactivation is established, it is irreversibly maintained, except in the germ cell line in which reactivation occurs at a specific stage of germ cell development. Therefore both X-chromosomes are functional throughout ovogenesis.
23. The inactive X-chromosome replicates in the latest part of S phase of cell cycle, when the autosomes and the active X-chromosome have already replicated.
24. The early support for Lyon's hypothesis stating that in female mammals dosage compensation occurs through inactivation of all but one X-chromosome, thus producing functional equivalence in both the sexes, came from studies of Glucose-6-phosphate dehydrogenase (G-6-PD) activity in cells. Enzymatic activity was shown to be equal in two sexes. Two alleles of G-6-PD locus, which produce electrophoretically distinct enzymes (F and S forms) were tested with heterozygous females. Isolated cells from several biopsies of skin from the same heterozygous person were cloned. Each clone contained either the F or S form but never both supporting the hypothesis.
25. Recent studies have been aimed at understanding the molecular mechanism of inactivation. It has been suggested that if the primary change is in DNA, this could be accomplished by an increased methylation of CG bases. Some experimental support for this hypothesis comes from work in which treatment with 5-azacytidine, which leads to hypomethylation, has produced reactivation of X-chromosome genes. This treatment however does not produce the reactivation of the whole X-chromosome.
26. The dosage compensation in Drosophila occurs by a different mechanism in which the X-chromosome of male is hyperactive and is responsible for dosage compensation. The hyperactivity of X does not depend on sex physiology but on X:A ratio of a cell in the same manner as the sex in Drosophila is determined by X:A ratio.

SEX-LINKED INHERITANCE:

1. Sex chromosomes (X and Y) are primarily concerned with the determination of sex but these chromosomes also carry some genes for other body characters. Such characters whose genes are localised on sex chromosomes follow sex during inheritance and are known as sex-linked characters. The genes governing the sex-linked characters are called as sex-linked genes and their mode of inheritance is known as sex-linked inheritance.
2. The concept of sex-linkage was first introduced by Thomas Hunt Morgan in 1910, while working on Drosophila melanogaster.
3. Sex linked inheritance can be classified into three types based upon the localization of the gene on the sex chromosome. They are (a) X-linked inheritance (b) Y-linked inheritance and (c) X-Y linked inheritance.
4. X and Y-chromosomes are partly homologous and partly non-homologous.
5. The genes which are localized in the non-homologous region of X-chromosomes are called as X-linked genes and the inheritance shown by such genes is called as X-linked inheritance. X-linked genes do not have any corresponding allele in Y-chromosome.
6. Y-linked type of sex-linked inheritance is shown by those genes which are localized in the non-homologous region of Y-chromosome and that have no alleles on X-chromosomes. Y-linked genes are commonly called as Holandric genes.
7. X-Y-linked inheritance is shown by those sex linked genes which are localized on homologous sections of X and Y-chromosomes. This type of inheritance is also called as incompletely sex-linked because through the genes are sex-linked they do not follow any sex during inheritance and are equally expressed in both sexes.

8. X-linked characters are expressed more commonly in males than in females, because males express the character in hemizygous condition.
9. The X-linked genes show a characteristic pattern of inheritance called as criss-cross inheritance where a X-linked trait is transmitted from male parent (P_1) to male grandchild (F_2) through female child (F_1).
10. T.H. Morgan while working on Drosophila had observed a white eyed male in the culture of wild red eyed flies. White eye is due to a mutation in the gene which is responsible for the red colour of the eye.
11. The white eyed male was mated with red eyed female and all the progeny in F_1 were red eyed, indicating that white eye mutation (w) is recessive to red eye colour (w^+). When F_1 flies mate freely, the red and white eyed flies appeared in the ratio of 3:1.
12. But all the white eyed flies in F_2 generation were only males. About half of the F_2 males had white eyes and the other half had red eyes but all the female flies had red eyes.
13. Morgan explained this pattern by associating the gene for white eye with the X-chromosome. The white eyed male fly had only one X-chromosome and a Y-chromosome that lacked most genes of X, hence the allele for white eyes was hemizygous in males and thus expressed. The mutant allele present in the X-chromosome of the original white eyed male was passed onto his daughters. Therefore, all the F_1 red eyed females were heterozygous (w^+w) and are called as carriers as they carry the allele to next generation. The F_2 males receive the X-chromosome from their carrier mothers. 50% of the males receive X-chromosome with w^+ alleles and develop red eye colour white the other 50% of them receive X-chromosome with w allele and develop white eyes.
14. In a reciprocal cross when a red eyed male was crossed to a white eyed female, the F_1 progeny were all red eyed females (carriers) and white eyed males. When these F_1 progeny were selfed, in F_2 red eyed females, red eyed males, white eyed females and white eyed males were obtained in 1:1:1:1 ratio.
15. The inheritance of white eye colour in Drosophila can be explained as follows:
    (a) Gene for white eye colour is located on nonhomologous region of X-chromosome and Y-chromosome is empty, carrying no allele for eye colour.
    (b) Males express the white eye colour in hemizygous condition white females express in homozygous recessive condition. Hence it is more common in males than in females.
    (c) Heterozygous females are red eyed and are called as carriers. Males are never carriers.
    (d) White eyed female always have white eyed father and always produce white eyed male progeny.
    (e) The white eyed males receive X-chromosome with w gene from their mother and Y-chromosome with no allele for it from father. Therefore X-linked characters are never transmitted directly from ♂ parent to male progeny.
16. X-linked inheritance is observed for about 200 traits in man. The most common examples are (a) Red-green colour blindness (b) haemophilia (c) congenital night blindness (d) Myopia (e) Juvenile glaucoma (hardening of eye ball) (f) optic atrophy (degeneration of optic nerve) (g) Duchenne muscular dystrophy (h) testicular feminization syndrome etc. All these disorders are due to X-linked recessive gene.
17. There are also few examples of X-linked dominant phenotypes in humans. They are hypophosphatemia (a type of vitamin D-resistant rickets) and defective tooth enamel. These characters are equally expressed in both males and females as they are due to a dominant gene.
18. Red-green colour blindness is due to a recessive gene. The retina of eyes of man contain certain colour sensitive cone shaped cells which are necessary for the distinction of red and green colours.

These colour sensitive retinal cells are formed and controlled by certain closely linked genes which remain located in X-chromosome. The individual having the recessive genes remain unable to distinguish red and green colours. Lack of chlorolabe pigment in the retinal cone cells results in inability to discriminate green colour and this defect is called as deuteronopia. Lack of erythrolabe pigment results in inability to discriminate red colour and this defect is called as protanopia.

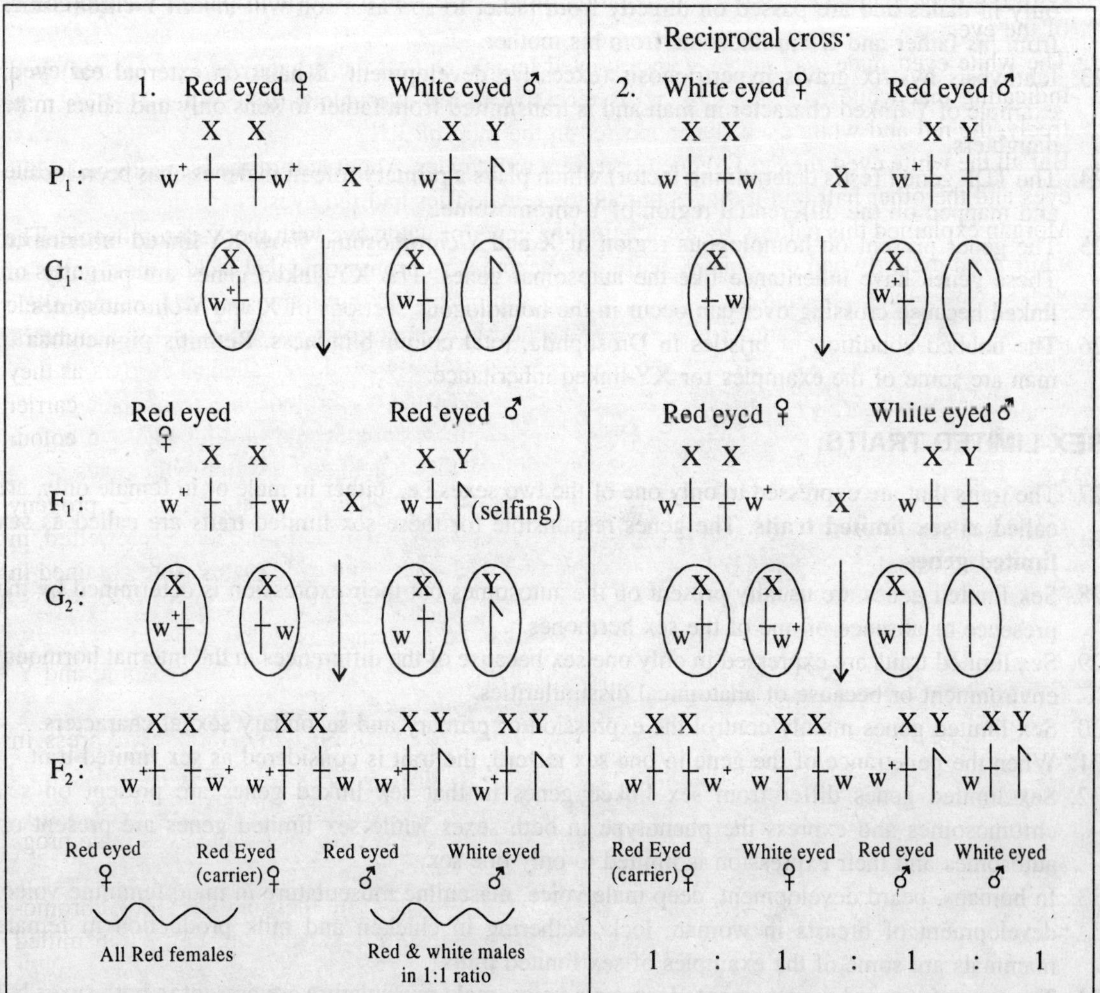

19. Color blindness is expressed in hemizygous condition in males and homozygous recessive condition in females. Therefore, the inheritance of colour blindness is similar to that of white eye in Drosophila.
20. Hemophilia is a bleeder's disease where the blood fails to clot when exposed to air and even a small skin injury results in continuous bleeding and can lead to death from loss of blood. In normal man, it takes 2-8 minutes to clot. A gene present on X-chromosome in its dominant form is necessary for the normal clotting process. It produces a necessary substrate thromboplastin for blood clotting. A mutant recessive form of this gene cannot produce thromboplastin, hence no

clotting takes place if the recessive gene is present in homozygous state in females and in hemizygous state in males.

21. Haemophilia is well known in the royal families of Europe where it is supposed to have come from Queen Victoria of Great Britain and inherited by her sons and grand sons through her heterozygous daughters.
22. The Holandric genes or Y-linked genes show Y-linked inheritance. The characters are expressed only in males and are passed on directly from father to son as a son will inherit Y-chromosome from his father and X-chromosome from his mother.
23. Ichthyosis hystrix gravis hypertrichosis (excessive development of hairs on external ear) is an example of Y-linked character in man and is transmitted from father to sons only and never to his daughters.
24. The TDF gene (Testis determining factor) which plays a primary role in maleness has been located and mapped on the differential region of Y-chromosome.
25. The genes present on homologous region of X and Y-chromosome show XY-linked inheritance. These genes have inheritance like the autosomal genes. The XY-linked genes are partially sex linked because crossing over can occur in the homologous sections of X and Y-chromosomes.
26. The bobbed condition of bristles in Drosophila, total colour blindness, Retinitis pigmentosa in man are some of the examples for XY-linked inheritance.

SEX-LIMITED TRAITS:

27. The traits that are expressed in only one of the two sexes i.e., either in male or in female only, are called as **sex limited traits**. The genes responsible for these sex limited traits are called as **sex limited genes**.
28. Sex limited genes are usually present on the autosomes but their expression is determined by the presence or absence of one of the sex hormones.
29. Sex limited traits are expressed in only one sex because of the differences in the internal hormonal environment or because of anatomical dissimilarities.
30. Sex limited genes mainly control the expression of primary and secondary sexual characters.
31. When the penetrance of the gene in one sex is zero, the trait is considered as sex limited trait.
32. Sex limited genes differ from sex linked genes in that sex linked genes are present on sex chromosomes and express the phenotype in both sexes while sex limited genes are present on autosomes and their expression is limited to only one sex.
33. In humans, beard development, deep male voice, masculine musculature in man, feminine voice, development of breasts in woman, lock feathering in chicken and milk production in female mammals are some of the examples of sex limited traits.
34. The genes for beard development, deep male voice, male musculature are present in both sexes but are expressed only in the presence of male sex hormone. Similarly genes for feminine voice, breast development are expressed only in the presence of female hormones.
35. A cow may receive genes for high milk production from its father (bull). Some bulls constantly produce female offsprings that show a higher milk yield than their mothers. Though the genes for milk yield are present in the bull, they are unable to express and produce milk in the absence of female hormones.

SEX INFLUENCED TRAITS:

36. Some genes are predominant in their expression of characters in one sex than in the other, due to

the differences in the internal environmental conditions provided by the sex hormones. Such genes are called as **sex influenced genes** and such traits as **sex influenced traits.**

37. Sex influenced genes reside on any of the autosomes or on homologous region of sex chromosomes.
38. Sex influenced traits are found in higher animals with well developed endocrine system.
39. Sex influenced genes behave as dominant in one sex and as recessive in the other.
40. Short index finger, pattern of baldness, gout in humans, horns in sheep are some of the examples of sex influenced traits.
41. The gene for pattern baldness is dominant in man and recessive in woman. Hence a man turns bald even if he receives only one gene for baldness while females turn bald only if they receive two genes for baldness. A single gene for baldness can express itself only in the presence of male hormones. The genotypes and phenotypes in males and females are as follows:

Genotype	Male phenotype	Female phenotype
$H^N H^N$	Normal	Normal
$H^N H^B$	Bald	Normal
$H^B H^B$	Bald	Bald

42. In sheep, the Dorest breed is homozygous for the horned gene and has horns in both sexes. The suffolk breed is homozygous for hornless gene and has horns in neither sex. When these two breeds of horned and hornless are crossed, the progeny were all horned males and hornless females. This clearly shows that the gene for horns acts as dominant in male and recessive in female.

EXERCISE:

1. What is sex linked inheritance? Explain the inheritance pattern of haemophilia in man.
2. What is sex determination? Name various mechanisms of sex determination. Give an account of genetically controlled sex-determination mechanism.
3. Write in detail about chromatin bodies and dosage compensation.
4. Write short notes on
   (a) Sex determination in plants.
   (b) Sex limited traits.
   (c) Sex influenced traits.

12

Linkage and Crossing Over

LINKAGE

The tendency of genes to stay together during inheritance and to retain their parental combination even in the offsprings because of the genes being located relatively close to each other, in the same chromosome, is called as *Linkage* and the genes located on the same chromosome and being inherited together are known as *linked genes.*

Mendel's experiments on Pisum sativam showed that the seven different characters are transmitted independent of each other. Due to this independent assortment of characters, 9:3:3:1 dihybrid F_2 Ratio is obtained. These characters are independently assorted because the allelic pairs are on different sets of chromosomes and Mendel could explain the nature of inheritance successfully. But usually in an organism, the number of chromosomes are few while the number of genes are numerous. Therefore each chromosome must contain many genes. The genes located on the same chromosome cannot assort independently but tend to inherit together. This phenomenon of the tendency of the genes to remain together during inheritance is called *linkage* and all those genes which are located on the same chromosome constitute a *linkage group.* The total number of linkage groups in an organism is equal to the number of chromosome pairs. For example, there are 4 linkage groups is Drosophila melanogaster, 7 in pisum and 23 in man. The genetic markers are said to be linked whenever over 50% of the gametes produced contain parental combination of genes and less than 50% of the gametes contain recombinants of the markers.

Though the existence of linkage was predicted earlier, the theory of linkage was first propounded by T.H. Morgan in 1911. He formulated the 'Chromosome theory of linkage'. It has the following characteristics.

1. Genes that show linkage are situated in the same chromosome.
2. Genes are arranged in a linear fashion in the chromosome i.e. linkage is linear.
3. The distance between the linked genes is inversely proportional to the strength of linkage. The genes which are closely located show strong linkage, whereas the genes which are widely separated show a weak linkage.
4. Linked genes remain in their original combination during the course of inheritance.

PHASES OF LINKAGE

Depending upon the arrangement of linked genes on homologous chromosomes, there are two different phases in linkage. They are (a) Coupling phase and (b) Repulsion phase. Bateson and Punnet (1906) had described these phases.

(a) Coupling phase

The tendency of the two dominant genes to remain together during gamete formation of F_1, when they come from the same parent, is called as *coupling phase.*

Bateson and Punnet had crossed a variety of sweet pea (Lathyrus odaratus) having purple flowers and long pollen, with another variety having red flowers and round pollen. The gene for purple (R) is dominant over Red (r) and the gene for long pollen (L) is dominant over round (*l*). The F_1 offsprings showed purple flowers with long pollen having genotype RrL*l*. The F_1 plants are test crossed. In F_2, four types of phenotypes i.e. purple long, purple round, red long and red round appeared in 7:1:1:7 ratio. If the genes assort independently the four phenotypes should appear in 1:1:1:1 ratio (equal proportion) but a deviation from this ratio shows the linkage of the genes. In the progeny, purple long and red round plants appear with high frequency. From this, they have concluded that there is a tendency for allels coming from the same parent to remain together rather than to separate. This was described as *coupling phase.*

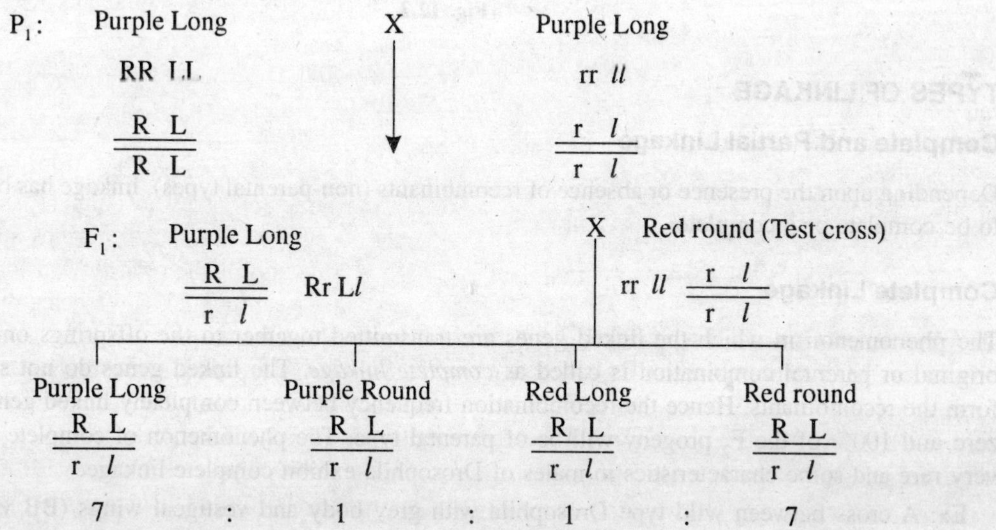

Fig. 12.1

(b) Repulsion phase

The tendency of the two dominant genes to separate out during gamete formation, when they come from two different parents is called as *Repulsive phase.*

Bateson and Punnet had crossed a plant having purple flowers and round pollen with a plant having Red flowers and long pollen. The F_1 offsprings showed purple round, Red long and red round plants appeared in 1:7:7:1. This ratio is obtained because F_1 forms gametes with BL or b*l* with lesser percentage than those with B*l* and bL genes.

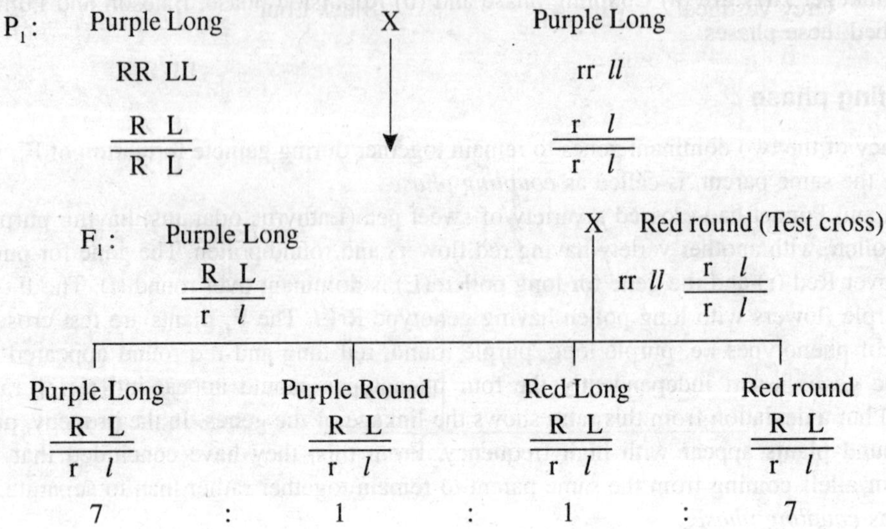

Fig. 12.2

TYPES OF LINKAGE

Complete and Partial Linkage

Depending upon the presence or absence of recombinants (non-parental types), linkage has been found to be complete or incomplete.

Complete Linkage

The phenomenon in which the linked genes are transmitted together to the offsprings only in their original or parental combination is called as *complete linkage.* The linked genes do not separate to form the recombinants. Hence the recombination frequency between completely linked genes will be zero and 100% of the F_2 progeny will be of parental type. The phenomenon of complete linkage is very rare and some characteristics in males of Drosophila exhibit complete linkage.

Ex: A cross between wild type Drosophila with grey body and vestigeal wings (BB vg vg) and black body long winged (bb Vg Vg) produces F_1 offsprings all of which have grey body and long wings (Bb Vg vg). These F_1 male hybrids, when test crossed with a double recessive female, produce offsprings of two types in equal proportion. This progeny resembles parental types (P_1).

The result indicates that grey body character is inherited together with Vestigeal wings and black body with normal wings. As all the progeny exhibits only the parental combination with no recombinants, the process is called as *complete linkage.*

Incomplete or Partial Linkage

The phenomenon in which some recombinants appear in the F_2 progency along with the parental types is known as *incomplete linkage.* The recombinants appear as a result of crossing over taking place between two homologous chromosomes. Incomplete linkage is very common and is observed in almost all organisms.

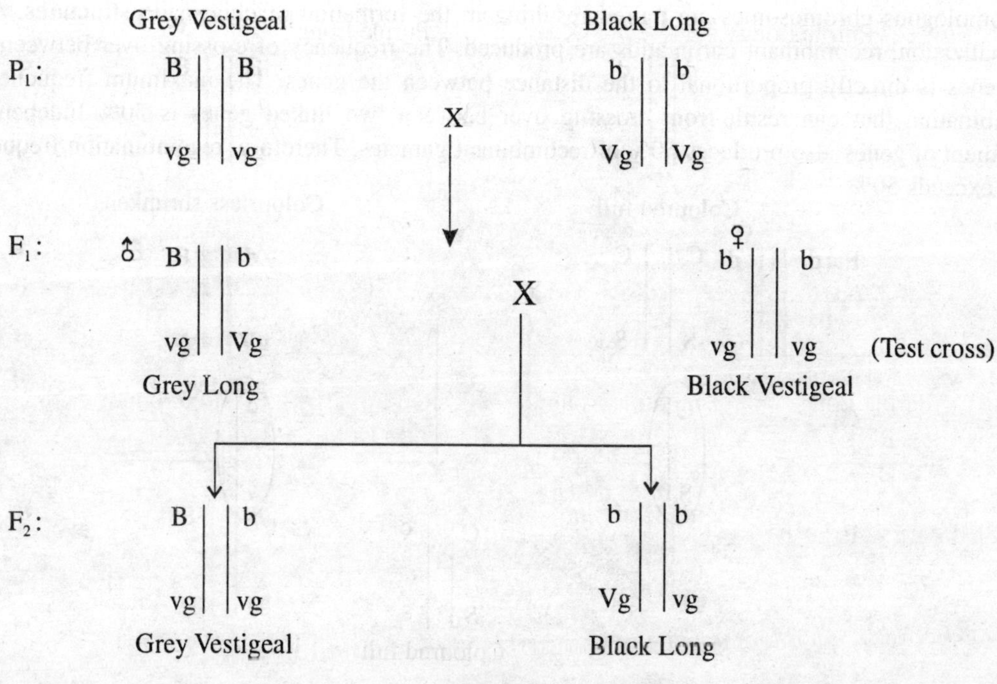

Fig. 12.3

Example

(1) **Incomplete linkage in maize:** Hutchinson crossed a maize variety of coloured and full endosperm seeds (CCSS) with another variety of colourless and shrunken seeds (ccss). All the F_1 plants produce coloured and full seeds. When F_1 plants are test crossed, four types of seeds are produced. These are:

(a) Coloured full – 4032/8368
(b) Colourless shrunken – 4035/8368 } 96.4%

(c) Coloured shrunken – 149/8368
(d) Colourless full – 052/8368 } 3.6%

If there is independent assortment of genes, the expected ratio is 1:1:1:1 and if there is complete linkage, only the parental combinations should appear in 1:1 ratio. As 3.6% of recombinants are observed along with 96.4% of parental combinations, the linkage is said to be incomplete or partial. These recombinants arise as a result of crossing over occuring between C and S genes.

CROSSING OVER

The process in which interchange (exchange) of chromosomal segments between non-sister chromatids of homologous chromosomes occurs resulting in the formation of recombinants is called as *crossing over*.

Crossing over and independent assortment are mechanisms that produce new combination of genes. Crossing over is the mechanism for the formation of recombinants for linked genes while the genes present on different chromosomes assort independently and form recombinants along with parental types. Crossing over in case of linked genes occurs in the pachytene stage where the paired homologous chromosomes show 4 chromatids and hence are called as tetrads. Crossing over takes place after the homologous chromosomes are paired resulting in the formation of chiasmatic structures. After terminalization, recombinant chromatids are produced. The frequency of crossing over between any two genes is directly proportional to the distance between the genes. The maximum frequency of recombination that can result from crossing over between two linked genes is 50% Independent assortment of genes also produces 50% of recombinant gametes. Therefore, recombination frequency never exceeds 50%.

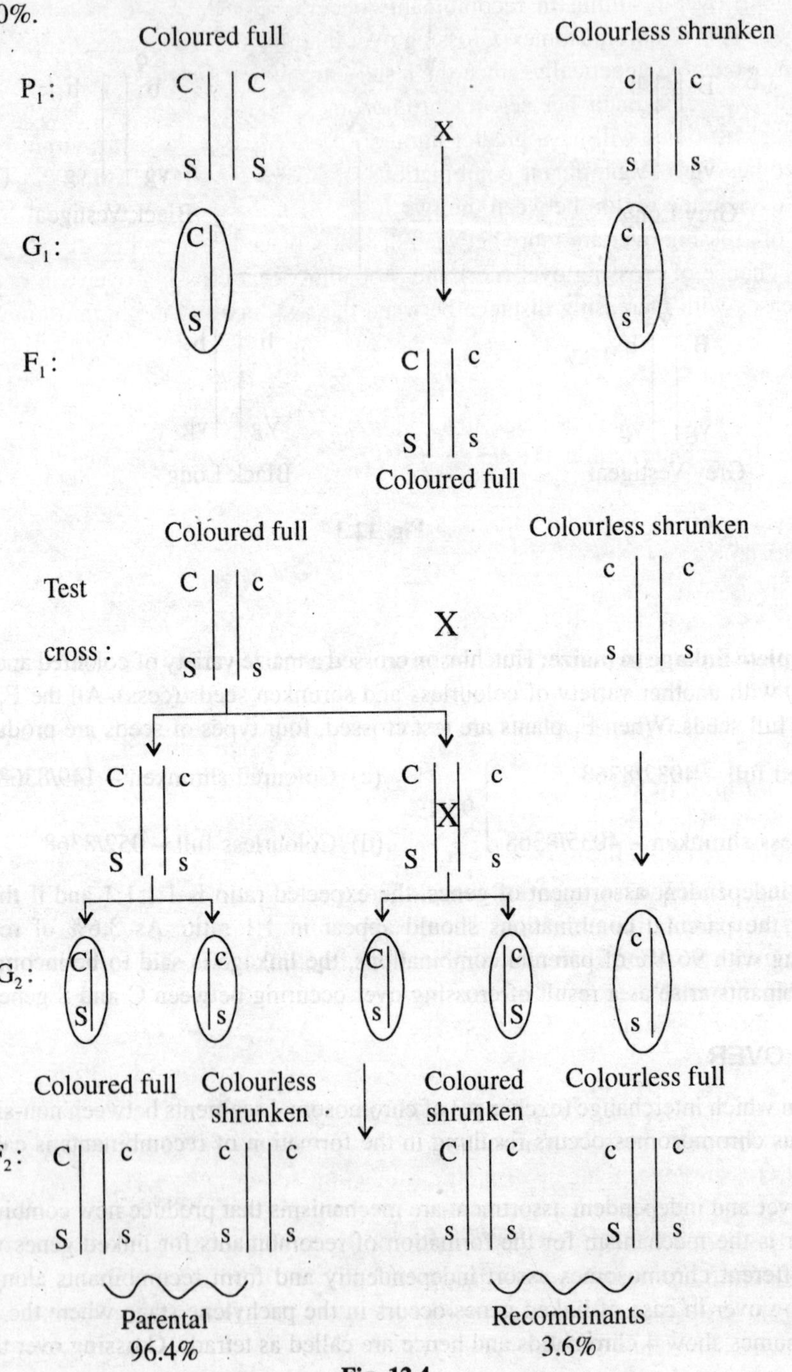

Fig. 12.4

Essential features of crossing over

1. Crossing over occurs at Pachytene stage after the synapsis of the homologous chromosomes has occured in prophase I to meiosis.
2. Only two of the four chromatids are involved in crossing over.
3. The crossing over resulting in recombinants occurs between the non-sister chromatids of the homologous pair of chromosomes. Crossing over that involves sister chromatids also occurs, but it is seldom detectable genetically, since the sister chromatids are genetically identical.
4. The number of chiasmata per set of chromosomes depends upon the length of chromosome. Longer chromosomes will have greater number of chiasmata than short chromosomes.
5. Chromosomes with recombinant combinations of linked genes are formed by the occurence of crossing over in the region between the two loci.
6. Chances of crossing over are more between distantly placed genes while closely linked genes will have less chance of crossing over (i.e.), the probability that crossing over will occur between two loci increases with increasing distance between the two loci on the chromosome.

Kinds of crossing over

There can be variable number of chiasmata along the length of a chromosome. The number of chiasmata usually depends upon the length of the chromosome. Depending upon the number of chiasmata, crossing over can be of the following types.

1. **Single cross over:** If only one chiasma is formed all along the length of chromosome pair, it is known as single cross over. The chromatids of homologous chromosomes contact and break only at one point.
2. **Double cross over (DCO):** In DCO, chromatids break and rejoin at two points i.e. two chiasmata are formed along the entire length of the chromosome. In D.C.O., the formation of each chiasma is independent of the other.

   Depending on the type of chromatids involved in D.C.O. three different types of D.C.O. are formed. They are

(a) **Two str and DCO:** In this type, two chiasmata are formed between the same two chromatids of homologous chromosomes. As a result, two parental and two crossover chromatids are formed.

(b) **Three str and DCO:** In this type, three str and s are involved in the crossing over. As a result, only one chromatid is of non-crossover type and three are crossover type.

(c) **Four str and DCO:** In such C.O., all the four chromatids of a homologous pair are involved in crossing over producing all crossover type chromatids. No chromatid is of non-crossover type.

(d) **Multiple cross over:** When crossing over occurs at more than two places in the same chromosome pair, it is known as multiple crossing over.

Mechanism of crossing over

Observations under microscope have disclosed the behaviour of chromosomes during meiosis and the mechanism of crossing over. This involves the following steps:

(a) Synapsis and formation of Synaptinemal complex:

During prophase-I of meiosis the maternal and paternal chromosomes of homologous pair come close together and pair at zygotene stage and lie side by side all along their lengths. The paired chromosomes are known as bivalents. The pairing of homologous chromosomes occurs due to the shortage of

DNA and histones in chromosomes and some amount of DNA and histone synthesis occurs during zygotene stage.

Once the homologous chromosomes pair, they form a complex structure called as synaptinemal complex, first described by Montrose J. Moses in 1955. It occurs as a highly organized structure of filaments between the paired chromosomes in zygotene and pachytene stages of meiosis. The electron micrographs of synaptinemal complex show three parallel dense lines equally spaced in one plane and flanked by chromatin. Some fine transverse str and s cross between lateral elements connecting them with the central element. The lateral elements may show subdivisions into longitudinal components. The central element may also appear as a long tripartite bar with ladder-like transverse connections. The synaptinemal complex contains DNA and protein and is believed to have no role in molecular paring of DNA str and s but may facilitate effective synapsis.

(b) Duplication of chromosomes

Each of the homologous chromosomes in a bivalent, splits logitudinally into two sister chromatids. Thus the bivalent consists of four chromatids and is known as a tetrad. The longitudinal splitting of chromosomes is achieved by the separation of already duplicated DNA molecules.

(c) Chiasmata formation

When the paired chromosomes start separating, the chromatids remain in contact at one or more points and thus establish one or more exchanges per bivalent and these points of contact are known as chiasmata. At each chiasma two non-sister chromatids of the bivalent break at the corresponding points and then rejoin with the exchange of segments. The breakage of chromatids is brought about by a nuclear enzyme, endonuclease and the fusion of broken segments takes place due to the action of enzyme ligase. A little amount of DNA synthesis (0.3% of the total genome) takes place during the crossing over that repairs the broken chromosome.

(d) Terminalization

After crossing over, the non-sister chromatids start repelling each other, because the forces at attraction keeping them together lapse. The chromatids separate from the centromere towards the tip and the chiasmata moves in zipper like fashion towards the ends. The movement of chiasmata is known as *terminalization.*

Cytological Basis of Crossing over

Morgan first proposed crossing over to explain the formation of recombinant combination of liked genes. He hypothesized that this linkage was due to the location of genes on the same chromosome. If crossing over occurs, one might expect to be able to observe it cytologically. Infact, cross shaped structures, in which two of the 4 chromatids of homologous chromosomes pairs appear to exchange the chromosomal segments, are readily detected in cytological studies of meiosis in many organisms. These cross shapeo structures were first detected by F. Janssens in amphibians and are called as chiasmata. A direct relationship between crossing over and chiasmata was observed when the chiasma frequencies were correlated with recombination frequencies.

Direct cytological evidence that homologous chromosomes exchange parts during crossing over was first obtained in 1931 by Curt Stern, working with Drosophila and by H.B. Creighton and B.Mc Clintock working with maize. Normally the two homologous of any pair are morphologically similar and are indistinguishable. However, Stern, Creighton and Mc Clintock identified homologues that were morphologically distinguishable i.e., they were not completely homologous. The chromosome pairs were homologous along most of their length such that they paired and segregated normally during meiosis. However, the homologs differed at their ends having distinct morphological features that could be recognized by microscopy.

1. Stern's Experiment in Drosophila

In 1931, Curt Stern demonstrated the mechanism of genetic crossing over under microscope. He studied two X chromosomes that were morphologically different from the normal X chromosome of Drosophila. One X chromosome had part of Y chromosome attached to it at one end and the second one was shorter than normal as a segment had been broken and translocated to chromosome IV. Female flies heterozygous for these two morphologically distinguishable X chromosomes were obtained and these flies were also heterozygous for alleles of two genes that are located on the X-chromosome. One gene affects eye shape and the other eye colour. The partially dominant mutant allele B results in bar shaped eyes while its wild type allele B^+ produces round eyes when homozygous. The mutant allele car of the second gene results in carnation eye colour and its dominant allele car^+ yields red eyes. The female flies in Stern's experiment carried the allelic pairs is Cis configuration as shown.

Stern test crossed such heterozygous flies with males having carnation coloured and normal shaped eyes having genotype car B^+/Y. The cross can be represented as follows:

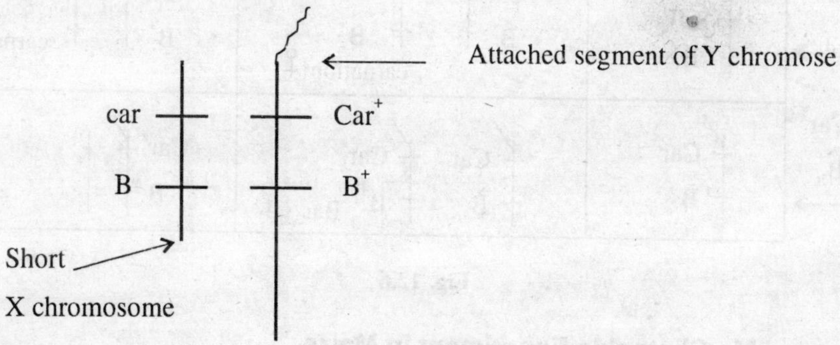

Fig. 12.5

Stern determined the genotypes of the progeny and observed whether the progeny with recombinant genotypes carried X-chromosomes with recombinant combinations of the morphological markers. The morphological markers present on the X chromosomes of each recombinant progeny were precisely those predicted if crossing over involved the breakage and exchange of parts of homologous chromosomes. Four phenotypic classes are obtained in the progeny. The male flies receive the X chromosome from the mother while the female receives one X from the paternal parent and other X from the maternal parent. When the X chromosome were observed cytologically, the maternal 'X' chromosome in the non crossover males and females carried either deletion or translocation while the recombinant progeny with bar shaped eyes carried the X chromosome with deletion at one end and also with translocated Y chromosome at other end. Other recombinant flies with carnation eye colour contained long X chromosome without deletion and translocation.

These recombinant flies carrying the new combination of genes are obtained if a crossover between car and B loci occurs in female fly and the morphological markers present at different ends are also exchanged, which can be observed cytologically.

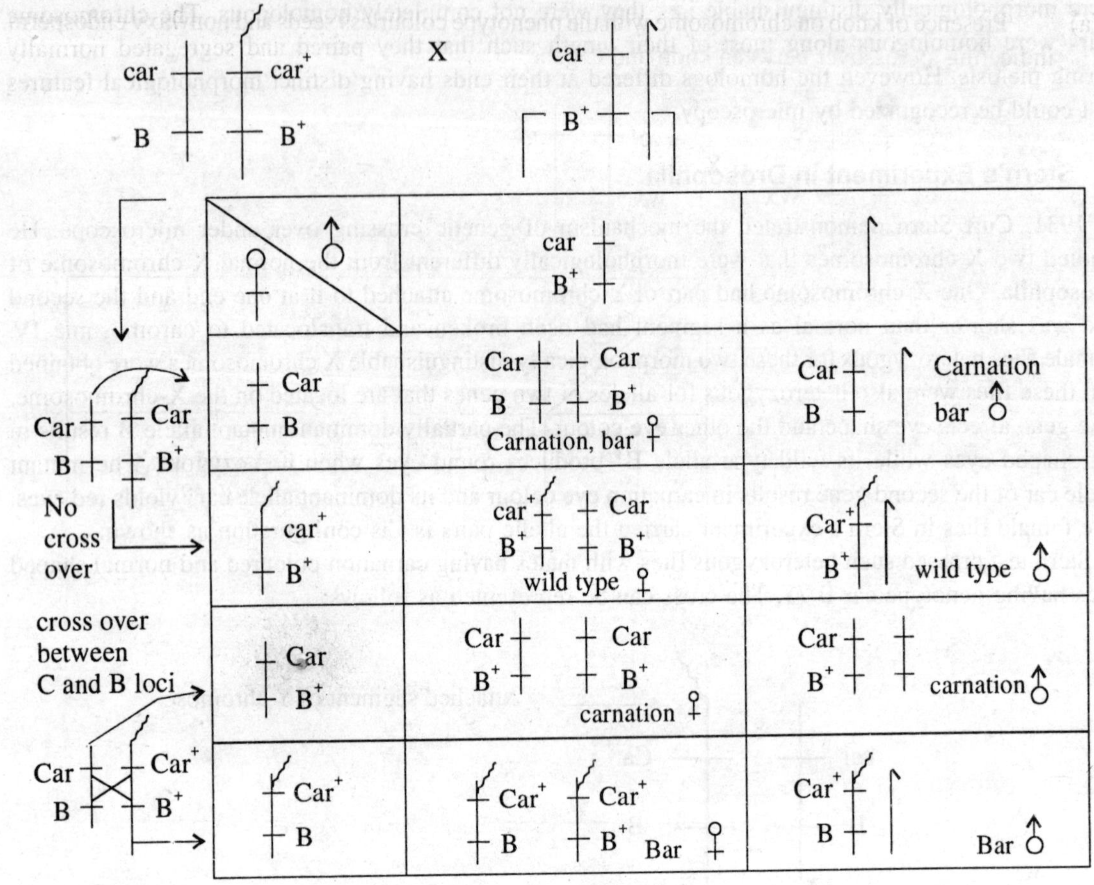

Fig. 12.6

2. Creighton and Mc Clintock's Experiment in Maize

Creighton and Mc Clintock (1931) obtained a corn plant in which 9th chromosome had a knob at one end and a segment of 8th chromosome at the other end. This plant was heterozygous for coloured aleurone and waxy endosperm and the genes are in repulsive phase i.e. C wx/c Wx. When this plant was test crossed, there will be two types of non crossover gametes and six types of crossover gametes and the progeny is of following types:

(1) Coloured waxy with knob	(Cwx/cwx)	Non crossovers
(2) Colourless nonwaxy without knob	(cWx/cwx)	
(3) Colourless nonwaxy with knob	(cWx/cwx)	Single cross over between knob and C gene
(4) Coloured waxy without knob	(Cwx/cwx)	
(5) Coloured nonwaxy with knob	(CWx/Cwx)	Single crossover, between C and Wx loci
(6) Coloured nonwaxy with knob	(cwx/cwx)	
(7) Colourless waxy with knob	(cwx/cwx)	Double crossover between C and knob-C and Wx.
(8) Coloured nonwaxy without knob	(CWx/cwx)	

The study of chromosomes of the offsprings with different phenotypes indicated that in all the crossover types genetic recombination has occurred based on the following observations.

(a) Presence of knob on chromosome with the phenotype colourless seeds and nonwaxy endosperm, indicating a crossover between knob and C gene.

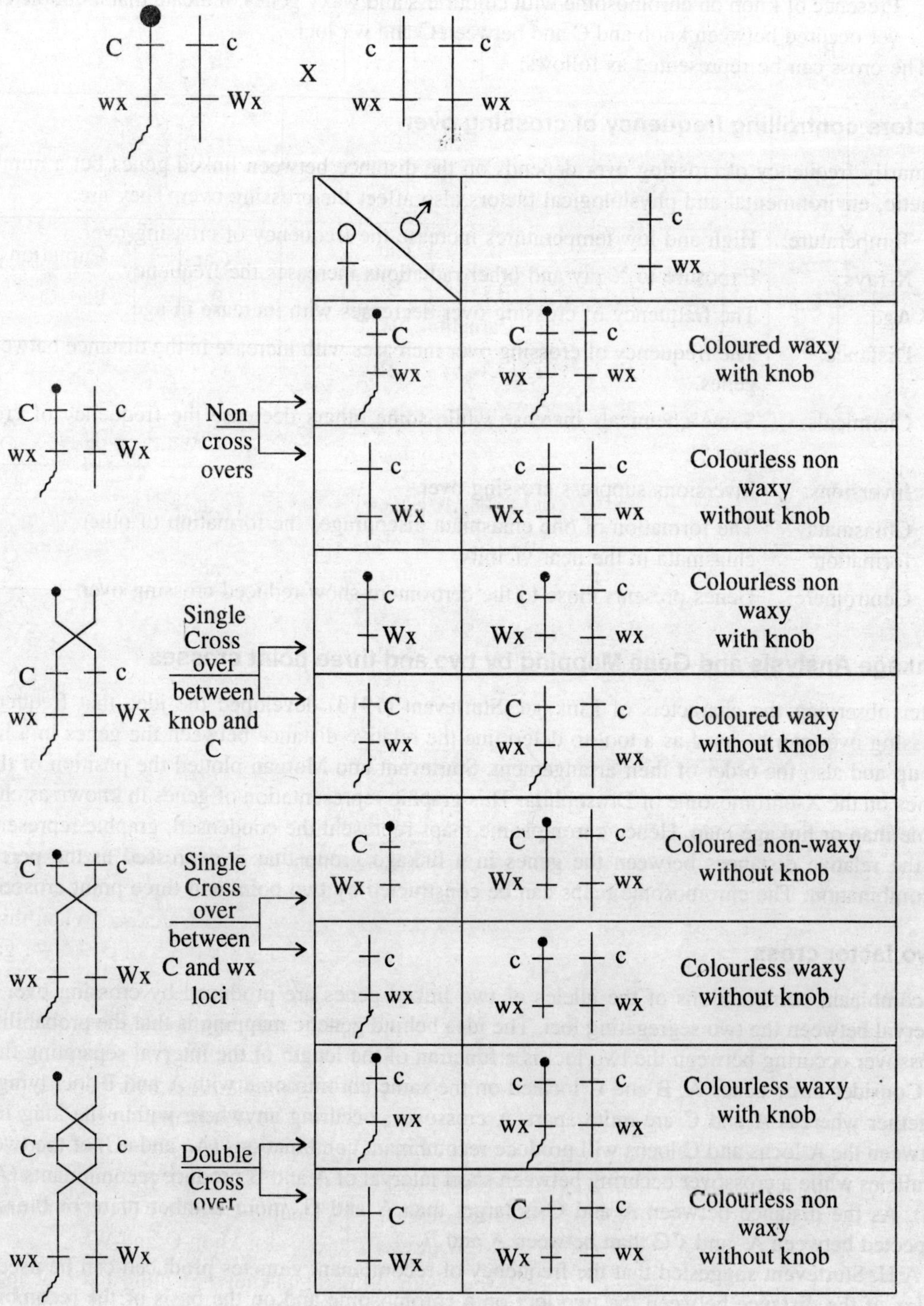

Fig. 12.7

(b)Presence of knob on chromosome with the phenotype coloured nonwaxy indicating a crossover between C and wx loci.

(c) Presence of knob on chromosome with colourless and waxy genes, indicate that a double crossover occured between knob and C and between C and wx loci.

The cross can be represented as follows:

Factors controlling frequency of crossing over

Primarily frequency of crossing over depends on the distance between linked genes but a number of genetic, environmental and physiological factors also affect the crossing over. They are:

1. Temperature: High and low temperatures increase the frequency of crossing over.
2. X-rays: Exposure to X-ray and other radiations increases the frequency.
3. Age: The frequency of crossing over decreases with increase in age.
4. Distance: The frequency of crossing over increases with increase in the distance between the genes.
5. Chemicals: Some chemicals increase while some others decrease the frequency of crossing over.
6. Inversions: Inversions suppress crossing over.
7. Chiasmata formation: The formation of one chiasmata discourages the formation of other chiasmata in the near vicinity.
8. Centromere: Genes presents close to the certromere show reduced crossing over.

Linkage Analysis and Gene Mapping by two and three point crosses

After observing the characters of Linkage, Sturtevant (1913) developed the idea that frequency of crossing over can be used as a tool to determine the relative distance between the genes in a linkage group and also the order of their arrangement. Sturtevant and Morgan plotted the position of the five genes on the X-chromosome of Drosophila. This graphic representation of genes in known as chromosome map or linkage map. Hence chromosome maps represent the condensed, graphic representation of the relative distances between the genes in a linkage group that is expressed as the percent of recombination. The chromosome maps can be constructed by two point and three point crosses.

Two factor cross

Recombinant combinations of the alleles of two linked genes are produced by crossing over in the interval between the two segregating loci. The idea behind genetic mapping is that the probability of a crossover occuring between the two loci is a function of the length of the interval separating the loci.

Consider three genes A, B and C located on the same chromosome with A and B loci lying close together whereas A and C are quite apart. A crossover occuring anywhere within the long interval between the A locus and C locus will produce recombinant combinations (Ac and aC) of the two pairs of alleles while a crossover occuring between short interval of A and B produce recombinants (Ab and aB). As the distance between A and C is larger than A and G, more number of recombinants are expected between A and s C than between A and B.

A.H. Sturtevent suggested that the frequency of recombinant gametes produced can be used as an index of the distance between the two loci on a chromosome and on the basis of the recombination frequency of the genes, linkage maps can be constructed. Linkage maps are made quantitative by defining one map unit as the distance that yields 1 percent of recombinants. Thus if A and B genes produce 5% of recombinants and A and C produce 10% of recombinants, then it indicates that A and B are 5 map units apart and A and C are 10 map units apart. As A, B genes are linked and A, C are also linked then B and C loci are also linked and the either 5 map units apart or 15 map units apart i.e., additivity can be achieved only by the following two linkage arrangements.

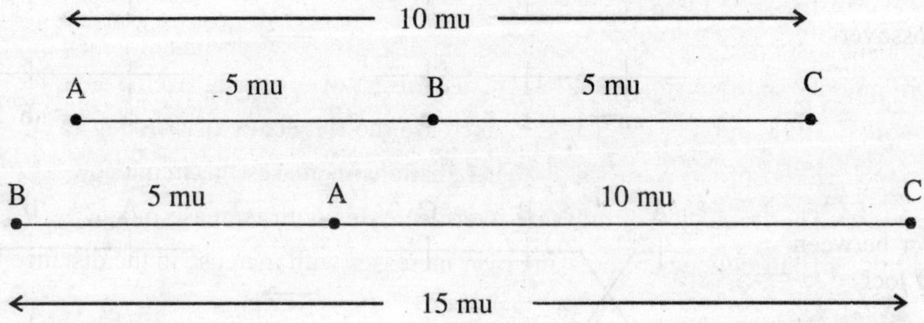

Fig. 12.8

But in actual crosses, when the distance between genes is large, the recombinants produced are significantly lower than the expected ones. This is because if two crossovers occur between the two genetic markers, it produces the parental combination of genes, hence double crossovers are not detected. For this reason, two factor crosses involving loci that are far apart will underestimate true linkage distances.

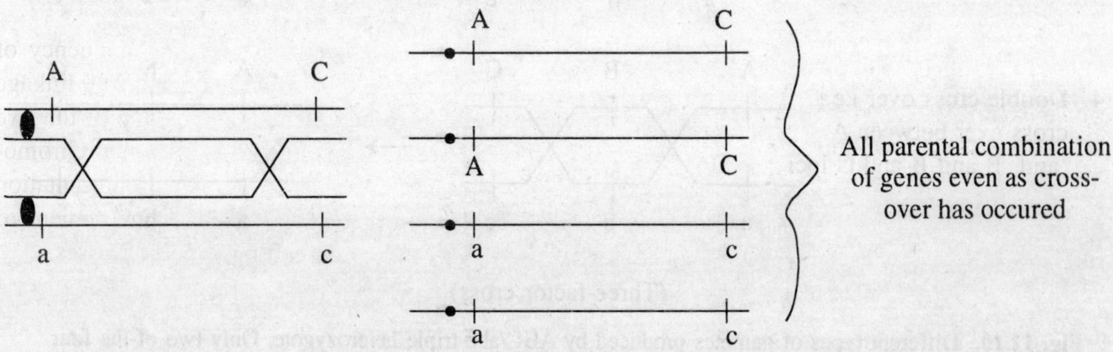

Fig. 12.9

Three Factor Crosses

A cross in which three pairs of alleles are segregating is known as a three-factor cross. A three factor cross allows the detection of double crossovers that are not recognizable in two factor crosses and also helps in finding the order of the markers involved. The three-factor cross is the most important tool used in chromosome mapping.

A cross is performed between two homozygotes to produce a triple heterozygote or trihibrid (ABC/abc) and the trihybrids are then test crossed to triple recessive individual so that the frequency of different F_1 gametes can be directly determined from the phenotypes of the test cross progeny. The trihybrids produce eight (2^3) different kinds of gametes. If all three genes assort independently, the eight gamete types occur with equal frequency. Deviations from this 1:1:l:1:1:1:1:1 ratio will occur if the genes are linked. The exact ratio will be determined by the degree of linkage between the loci involved.

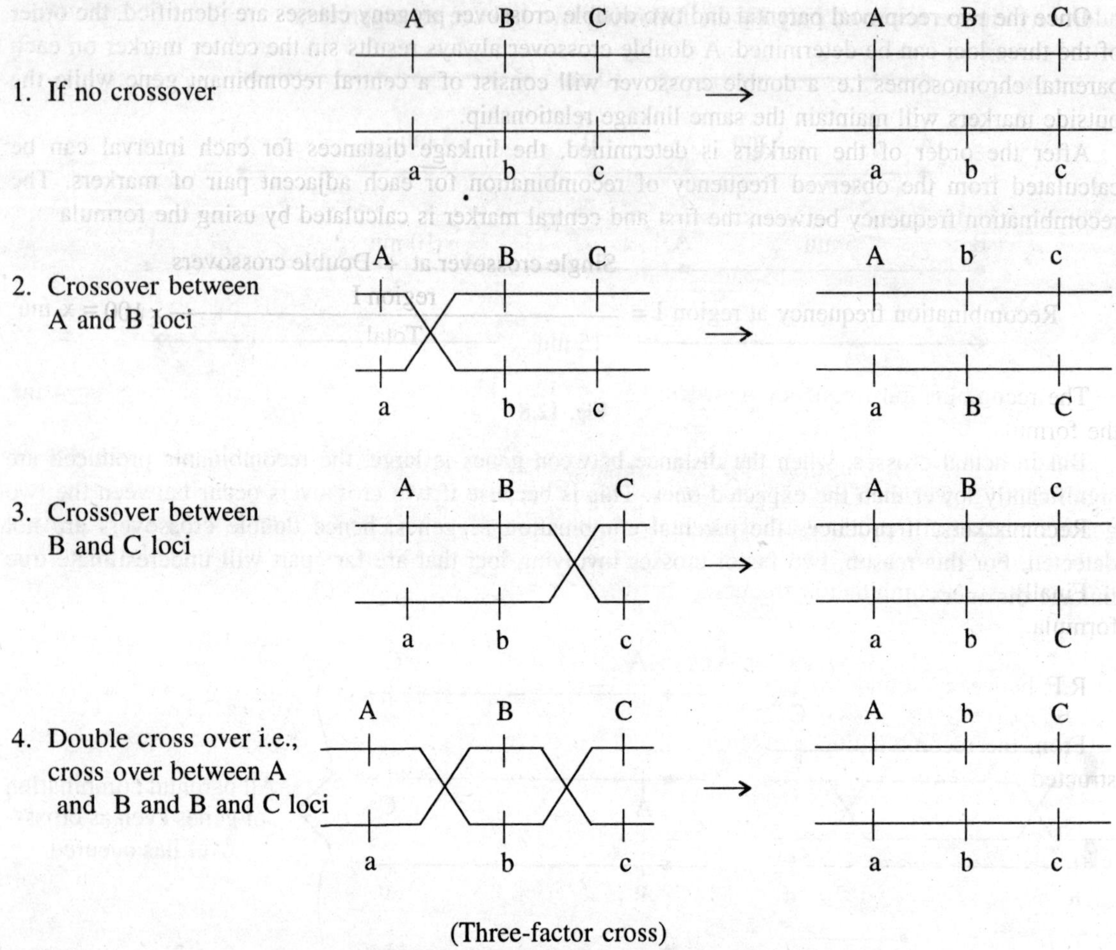

Fig. 12.10. Different types of gametes produced by ABC/abc triple heterozygote. Only two of the four chromatids involved in crossover are shown.

In three factor crosses, the parental type gametes will be present in highest frequencies. Hence the parental combinations can always be determined by identifying the two classes of test cross progeny that occur with the highest frequencies. The two parental progeny classes are expected to occur in approximately equal frequencies. The ability to identify the parental progeny classes can determine the linkage relationships of the genetic markers involved, that is whether the alleles of each pair of genes are presents in the cis or trans configuration on the homologous chromosomes of the F_1 progeny.

The two reciprocal double crossover progeny classes can also be identified. Among and eight phenotypic classes, there will always be two phenotypic classes which exist with least frequency indicating the double crossover types. If the two crossovers involved in a double crossover are independent events then the probability of a crossover occuring will be equal to the product of the probabilities of the two single crossovers occuring i.e. if single crossover at region I occurs with a probability of 'p' and single crossover at region II occurs with a frequency of 'q' then double crossovers occur with a frequency of 'pxq'. Since p and q are fractions less than 1/2, pq should necessarily be less than p or q. Hence double crossovers exist with least frequencies.

Once the two reciprocal parental and two double crossover progeny classes are identified, the order of the three loci can be determined. A double crossover always results sin the center marker on each parental chromosomes i.e. a double crossover will consist of a central recombinant gene while the outside markers will maintain the same linkage relationship.

After the order of the markers is determined, the linkage distances for each interval can be calculated from the observed frequency of recombination for each adjacent pair of markers. The recombination frequency between the first and central marker is calculated by using the formula

$$\text{Recombination frequency at region I} = \frac{\text{Single crossover at region I} + \text{Double crossovers}}{\text{Total}} \times 100 = x \text{ mu}$$

The recombination frequency between the central marker and third marker is calculated by using the formula.

$$\text{Recombination frequency at region II} = \frac{\text{Single crossover at region II} + \text{Double crossovers}}{\text{Total}} \times 100 = y \text{ mu}$$

Finally the recombination frequency between the two outside markers is calculated by using the formula.

$$\text{R.F. between 1st and 3rd loci} = \frac{\text{SCO I} + \text{SCO II}}{\text{Total}} \times 100 = z \text{ mu.}$$

From the recombination frequency values obtained by the above formula, a linkage map is constructed.

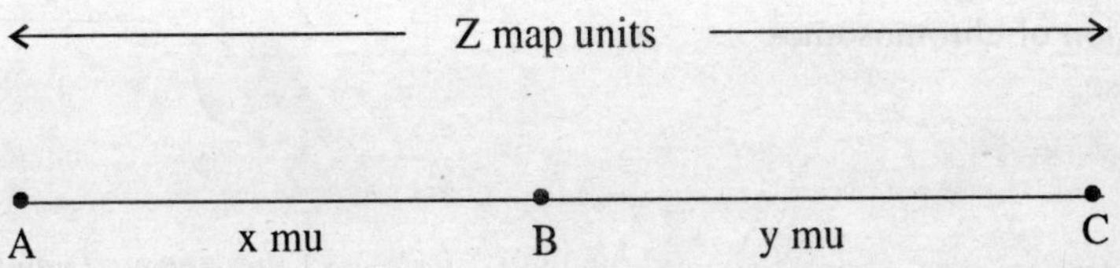

Interference:

In a three factor test cross, generally the observed number of double crossovers will be less than the expected double crossover. This is because the crossover formation at one region interferes with the crossing over and chiasma formation at other points nearby. This phenomenon is called as *chromosome interference* or *chiasma interference* or *Interference.* It was first observed by H.J. Muller in 1916. *Interference* is observed in most of the three factor crosses.

Although the molecular basis of interference is unknown, the observed levels of interference indicate that the occurence of one crossover decreases the likelihood of another crossover occuring nearby. The degree in interference is measured by the coefficient of coincidence which is the ratio of observed number of double crossovers to the expected number of double crossovers.

$$\text{Coefficient of coincidence} = \frac{\text{Observed frequency of Double crossovers}}{\text{Expected frequency of Double crossovers}}$$

where expected frequency of double crossover is calculated considering the two single cross overs to be independent events (pxq). In the absence of interference scans be calculated as one minus coefficient of coincidence.

Interference = 1 – Coefficient of coincidence

Interference values between 0 and 1 indicate that + ve interference had occured while the interference values between 0 and –1 indicate –ve interference i.e., the occurrence of one crossover increases the likelihood of the additional crossover occuring nearby. Negative interference is generally observed in bacteriophages. If the coefficient of coincidence is equal to 1 i.e., expected frequency of double crossover is equal to the observed frequency of double crossover, the interference value is zero indicating that there is no interference.

EXERCISE:

1. What is linkage? Explain in detail as to how it is correlated with Mendel's laws of independent assortment.
2. What is linkage? Describe the mechanism of linkage with suitable examples.
3. What is linkage? Explain different phases of linkage.
4. Give an account of linkage in Drosophila melanogaster.
5. Describe the mechanism and significance of crossing over.
6. Give an account of linkage in maize.
7. Write short notes on
   (a) Chiasma
   (b) Crossing over
   (c) Coupling phase
   (d) Repulsion phase
   (e) Complete linkage
   (f) Incomplete linkage
   (g) Duplication of chromosomes
   (h) Interphase.

13

Chromosomal Variations

Structural changes within chromosomes also known as chromosomal aberrations are transmitted through mitosis and meiosis which cause changes in phenotypes and in the expected genetic ratios. These changes are collectively called variations in chromosome structure. Structural variations in chromosomes are of four distinct types:

Deletions
Duplications,
Translocations, and
Inversions.

Chromosomal arrangements occur spontaneously due to various natural causes. Whereas, the induced chromosomal variations occur due to physical and chemical mutagens. These chromosomal variations may be confined to single chromosome or may be extended to both chromosomes of a pair or may involve two or more pairs of chromosomes. Thus it is of two kinds (1) Intrachromosomal variation and (2) Interchromosomal variation.

(1) Intrachromosomal variation:

The intrachromosomal variations or aberrations occur in one and the same chromosome; they are of three types viz.,

Deficiency or deletion
Duplication and
Inversion

(a) Deficiency or deletion: A deficiency means deletion or losing of a small part of a chromosome leading to loss of one or more genes. Deficiency arises from breakage occurring at random in both chromatids of a chromosome or only in one chromatid. The deletion may be caused by various sources such as radiation, magnetic effect, chemicals, drugs or viruses at any time during the cell cycle, either in somatic or in germ cells. Based on its location, a break or deletion may be terminal, when a single break occurs near the end of the chromosome; or interstitial when two breaks take place in the middle portion of the chromosome.

Every deletion or break produces two free ends which may behave in either of the following ways:

(i) Both the broken ends may be reunited so that original chromosomal structure is restored. This process is called restitution;
(ii) The broken ends may not unite thus forming a chromosomal segment without a centromere which is subsequently lost during cell division.

(iii) Presuming that two single breaks occur in two different chromosomes in a cell , and the deleted segment of only one chromosome is united with the raw broken end on the other chromosome, this is called exchange union.

Deletion of any portion of chromosome is usually lethal to a diploid organism due to its genetic imbalance.

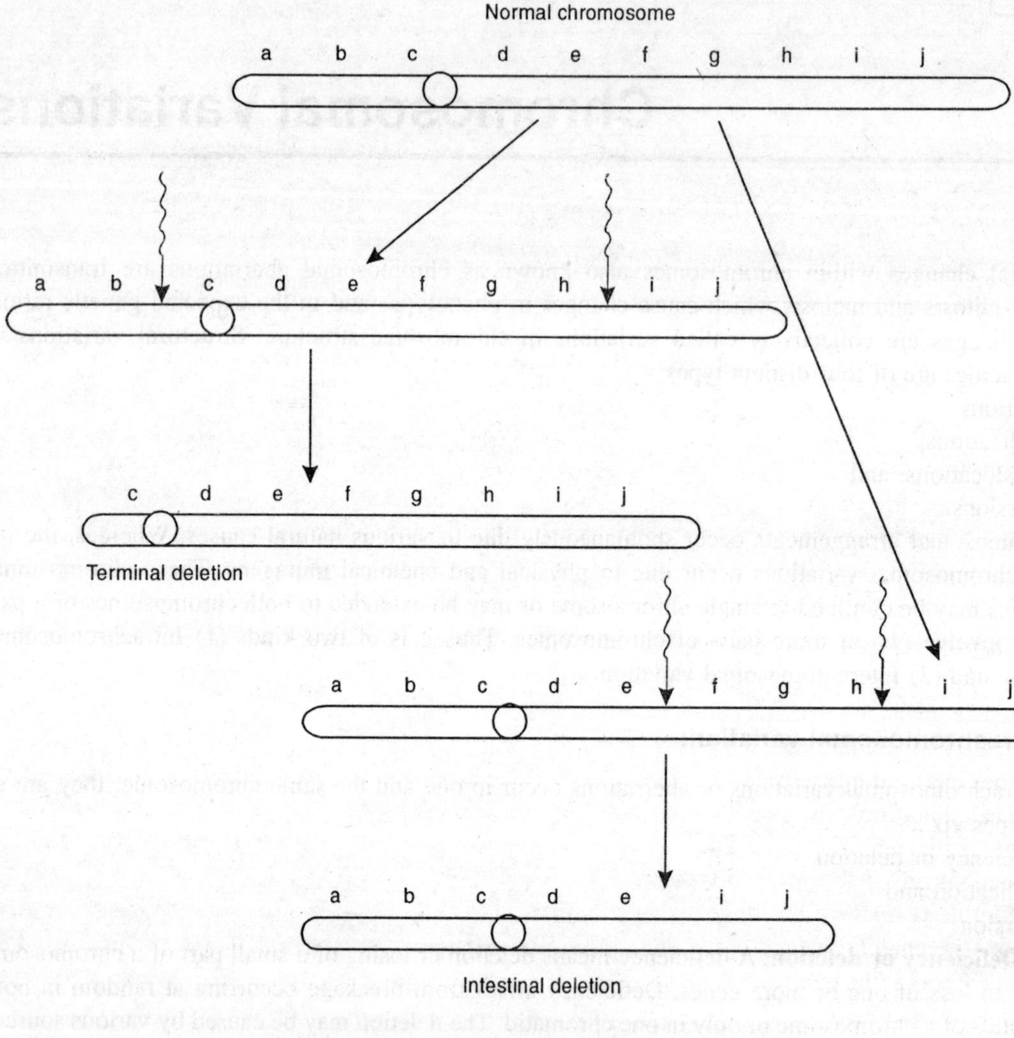

Fig. 13.1. Production of terminal and interstitial deletion

Fate of deleted fragment:

As the fragment does not possess a centromere, at metaphase it will not be able to get attached to spindle fibres and move towards a pole. It will remain at the centre of the cell and will not be included within, any of the two daughter nuclei.

It remains freely in the cytoplasm and will gradually disappear. Thus, the cell will lose one or few genes contained in the deleted chromosomal fragment.

(b) Duplication: Duplication involves attachment of a chromosomal fragment leading to addition of one or few genes to a chromosome. Automatically when there is a duplication in a chromosome, there shall be a corresponding deletion in another chromosome. Following illustrated types of duplications are known.

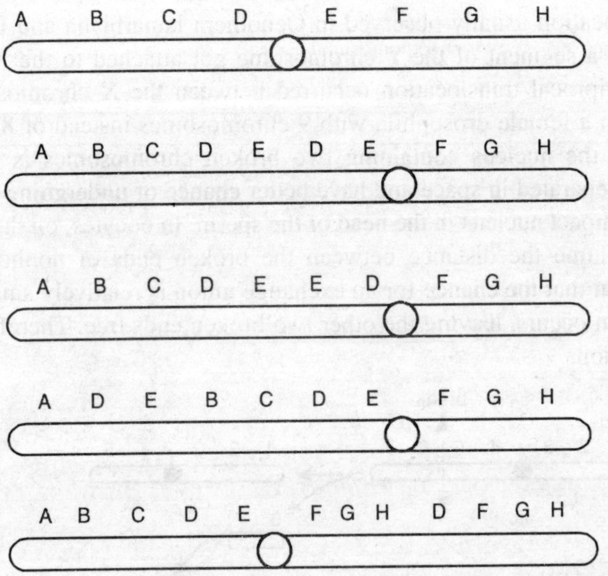

Fig. 13.2. Diagrams showing different types of duplications.

A gamete that receives a chromosome with a duplication will be diploid for few genes. When it fertilises a normal gamete, that zygote will have three sets of those genes that are present in the duplicated segment.

In female Drosophila, bar eyes is a dominant X-linked trait, which provides a range of interesting phenotypes due to duplication.

(c) Translocations: A segment of one chromosome gets detached and unites with another nonhomologous chromosome. Such an interchromosomal rearrangement is called translocation. Translocations are of following 5 types:

(i) **Simple translocation:** only a single break occurs in a chromosome and the broken fragment gets attached to the end of another chromosome. However, due to the presence of "nonsticky" telomeres at the unbroken ends of a chromosome, such a terminal attachment of a segment does not take place.

(ii) **Shifts:** In such translocation three breaks are involved. Two breaks occur in a chromosome to produce an interstitial fragment. This fragment gets inserted into one of the arms of another non homologous chromosome in which a single break has led to two "sticky" ends.

(iii) **Reciprocal translocations:** In this a single break occurs in each of the two nonhomologous chromosomes followed by a mutual exchange of the broken fragments resulting in two new chromosomes each having one segment of the other chromosome. Rarely, two breaks occur in each of the two chromosomes followed by exchange of intercatary segments. If the centromere containing segment of one chromosome is joined to the acentric piece of the other nonhomologue, the exchange is called eucentric. But if two centric pieces from two nonhomologues join to form a dicentric chromosome, it is called aneucentric. In the next division the dicentric chromosome will form a bridge and the acentric fragment will be lost. Therefore an aneucentric exchange unions are usually lethal. The eucentric reciprocal translocations produce viable gametes, provided both pairs of nonhomologous chromosomes exchange segments.

(iv) **Multiple translocations:** Sometimes more than two pairs of nonhomologous chromosomes may be involved in a translocation usually observed in Oenothera lamarbiana and Drosophila melanogaster. In Drosophila a segment of the Y chromosome got attached to the X chromosome. Simultaneously, a reciprocal translocation occurred between the X chromosome and chromosome IV, resulting in a female drosophila with 9 chromosomes instead of 8.

(v) **Half translocations:** When the nucleus containing two broken chromosomes is small, the broken ends are not widely separated in space and have better chance of undergoing reciprocal exchange. This is true for compact nucleus in the head of the sperm. In oocytes, on the contrary, due to the large nuclear volume the distance between the broken ends of nonhomologous chromosomes may be so great that the chance for an exchange union is relatively small. In that case only one exchange union occurs, leaving the other two broken ends free. Therefore this is considered as half translocation.

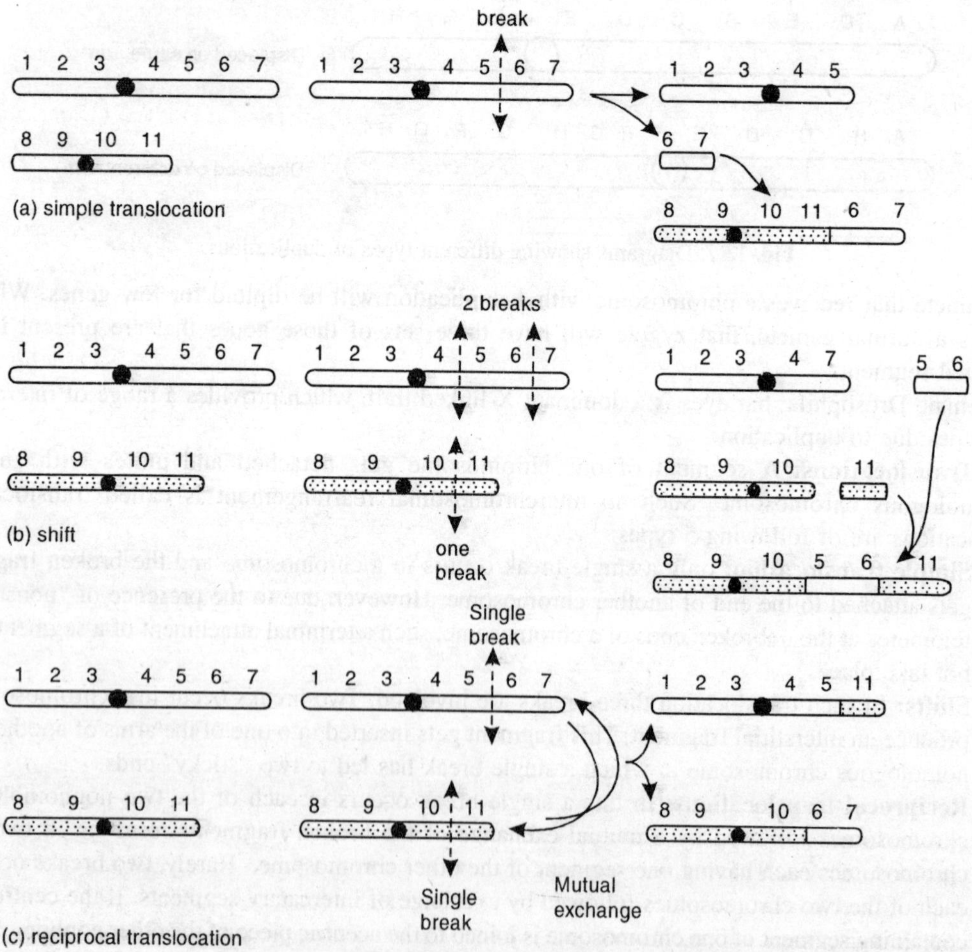

Fig. 13.3. Showing various types of translocations.

(d) Inversions: Inversions result when there are two breaks in a chromosome and the detached segment becomes reinserted in the reverse order. Depending upon the inclusion or absence of the centromere within the inverted segment, inversions are classified into two types viz., Paracentric inversion and Pericentric inversion.

Paracentric inversion: This type of inversion is identified in the heterozygote by formation of a pairing loop at pachetene. If the size of the loop is large enough, chiasma formation will take place within it. When a single chiasma forms between an inverted and a normal segment, the two chromatids involved will produce one dicentric chromatid and one acentric fragment after the exchange. However, the other two chromosomes will behave normal. Fig. 13.4.

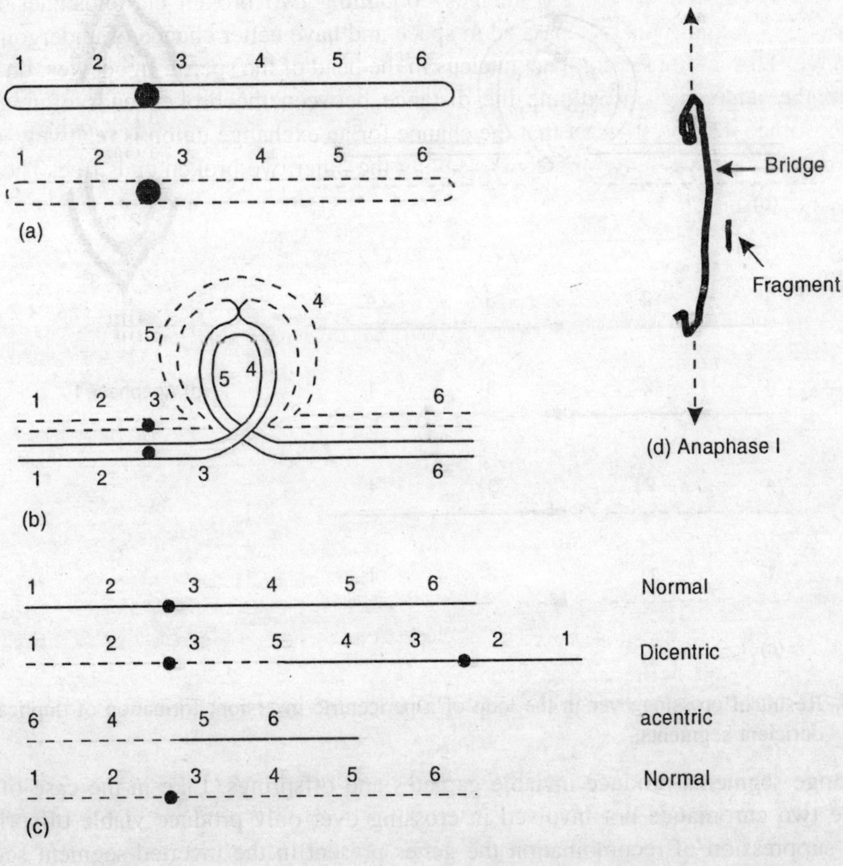

Fig. 13.4. Result of crossing over within the loop of a paracentric inversion. A chromosome bridge due to a dicentric chromosome and an acentric fragment are formed.

The dicentric chromosome at anaphase I will be pulled towards both poles; thus it will form a bridge that weakens and ultimately breaks. Eventually, the acentric fragment due to its inability to move would be lost. Subsequently, among the four gametes, two would be normal and two deficient in chromosome segments. Deficient plant gametes thus formed, are not viable. Whereas in animals such gametes participate in fertilisation but either zygote or embryo aborts. Paracentric inversions in this way suppresses recombination.

Pericentric inversions: In a heterozygous individual, the centromere is present within the loop for a pericentric inversion. The chromatids resulting after exchange do not form a dicentric and acentric fragment as in a paracentric inversion heterozygote, during chiasma formation. Instead, they have one centromere each, but are deficient for some segments, whereas other segments are duplicated. (Fig. 13.5)

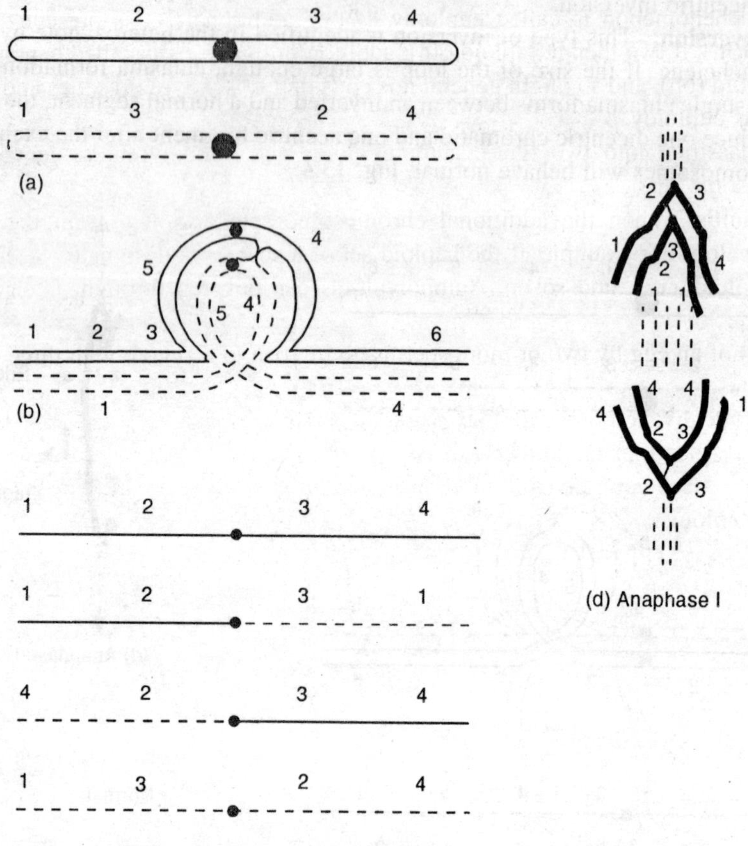

Fig. 13.5. Result of crossing over in the loop of a pericentric inversion; formation of duplication and deficient segments.

The exchange segments produce inviable gametes and offsprings. Like in the case of paracentric inversion, the two chromatids not involved in crossing over only produce viable offsprings. Subsequent to the suppression of recombination the genes present in the inverted segment segregate as a single unit called supergene within a population.

Numerical aberrations:

During the life time of an organism variations of many types occur causing useful, harmful or some interesting alterations in the genome. Any change in the genetic material which produces an altered phenotype are called mutations. The mutations involving cytologically visible changes in chromosomes are called chromosomal variations or aberrations. If the change occurs in a gene, it is referred to as gene or point mutation.

Numerical changes leading to increase or decrease in chromosome numbers are known as Polyploidy. When there is addition of one or many entire sets of chromosomes (a set means all chromosomes of a haploid complement), the condition is known as euploidy. Similarly, when one or more single chromosomes are added or deleted it is known as aneuploidy.

Euploidy: If an organism acquires additional set of chromosomes over and above the diploid complement; this phenomenon is called euploidy. If one additional set is present the condition is known as triploid (3n); if two then tetraploid (4n); addition of three sets is called pentaploid (5n), and of four sets hexaploid (6n), and so on. In general it is considered as polyploid series and the individuals are said to express euploidy.

Euploids are classified into two types, autopolyploids and allopolyploids, depending upon the additional chromosome set.

(a) Autopolyploids: When the additional chromosome sets originate from the same species, autopolyploids develop. For example, if the haploid set of a species is designated *x*, the diploid is *xx*, triploid xxx, tetraploid *xxxx*, and so on. Autopolyploids can develop through one of the following ways:

(i) Fertilisation of an egg by two or more sperms giving rise to a zygote with three or more sets of chromosomes.

(ii) Normal mitotic division in the diploid zygote in which chromosomes duplicate but cell division fails to occur, thus four haploid sets of chromosomes produce a tetraploid nucleus.

(iii) Due to failure of meiotic division in germ mother cells-unreduced diploid gametes are formed instead of haploids.

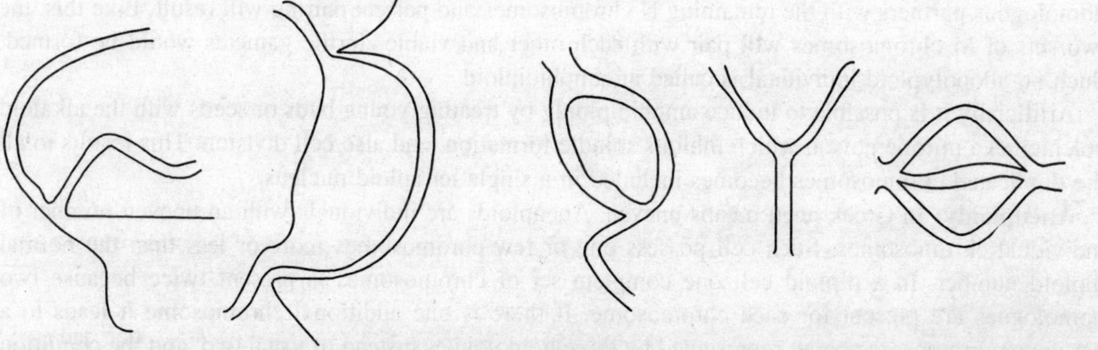

Fig. 13.6. Trivalent configurations in meiosis.

If three homologous are present they may or may not become paired to form a trivalent (Fig. 13.6). It is noteworthy that most economically popular varieties of seedless watermelons, bananas, European pears and apples including Indian carpet grass are triploid. They are developed from fertilisation between diploid gametes from tetraploid plants and haploid gametes from diploid plants. The triploid plants are healthy and robust and are propagated through asexual cuttings.

Polyploidy is more common in plants than in animals. Around 60 percent of angiosperms are known to be polyploids. Because plants are mostly bisexual (Hermaphrodite) organisms in which sex chromosomes do not play a significant role in normal growth and development. Increase in the number of chromosomes is therefore desirable as it increases phenotypic variability and magnifies the expression of some favourable traits. On the other hand in animals, polyploidy leads to a disturbance in the balance between sex chromosomes and autosomes. Increase in the number of sex chromosomes markedly affects sexual development. Due to this, polyploidy in animals is restricted to only hermaphrodites like leaches, earthworms and those which develop parthenogentically as in shrimps, aphids including some lizards. Another reason why polyploidy is more prevalent in plants is that the problem of sterility is easily overcome through asexual methods of reproduction viz. grafting and budding in plants. Such techniques are obviously not applicable to animals except that individual polyploid cells can be excised and cultures in the laboratory.

Evolution point of view polyploidy has played a significant role in evolution of plant species. Origin of certain important crop plants viz., barley, potatoes, grass (Dactylis glomerata), lotus and several ornamental plants is due to polyploidy.

Allopolyploidy: Allopolyploidy is the second type of euploidy where the additional set of chromosomes come from a different species. For example, a diploid species with 2 chromosome sets NN crosses with another species MM. The offspring produced would be NM which is viable but sterile. This is due to the chromosomes belonging to the set N do not find homologous partners in chromosomes of M, during meiosis. Due to the failure of pairing at anaphase I, the chromosomes move at random towards the two poles. Consequently each gamete gets an unbalanced mixture of N and M chromosomes and sterility results.

Of course, only one way of restoring fertility to a sterile hybrid (NM) is possible. During mitotic division in the NM hybrid all the chromosomes are allowed to divide but cell division has to be inhibited, the result would be a tetraploid nucleus with two sets of N and two sets of M chromosomes (NNMM). Because, when meiosis starts, all chromosomes belonging to one set of N will find homologous partners with the remaining N chromosomes and perfect pairing will result. Like this, the two sets of M chromosomes will pair with each other and viable fertile gametes would be formed. Such an allopolyploid individual is called an amphidiploid.

Artificially it is possible to induce amphidiploidy by treating young buds or seeds with the alkaloid colchicine, a mitotic poison which inhibits spindle formation, and also cell division. This results to all the duplicated chromosomes becomes included in a single tetraploid nucleus.

Aneuploidy: In Greek aneu means uneven. Aneuploids are individuals with an uneven number of individual chromosomes. Such cell possess one or few chromosomes more or less than the normal diploid number. In a diploid cell one complete set of chromosomes is present twice because two homologues are present for each chromosome. If there is one additional chromosome it leads to a particular chromosome being represented by three homologues instead of usual two, and the condition is known as trisomy ($2x + 1$). If there are two additional chromosomes in a cell so that one chromosome being represented by four homologues, it is called tetrasomy ($2x + 2$). It may also happen that the two additional chromosomes are homologues belonging to two different chromosomes of the complement so that two chromosomes are represented each by three homologues. Such an individual is said to be trisomic for two different chromosomes ($2x + 1 + 1$).

A different situation also exists involving loss of a chromosome from a diploid complement. This results in the presence of one chromosome without a homologous partner called monosomy ($2x - 1$). If both homologues of a chromosome are missing, it is called nullisomy ($2x - 2$).

Aneuploidy results from an abnormal incident of nondisfunction of single chromosome during meiosis. The paired homologues may fail to separate at metaphase so that both homologues reach same pole. Out of the four gametes formed from such a cell, two will have an extra chromosome each (trisomy) in addition to the haploid number ($x + 1$). The other two gametes will possess one less (nullisomy) than the haploid number ($x - 1$). If fertilisation takes place with a normal haploid gamete (n) from the other parent, then these four gametes will produce two types of aneuploid individuals –

trisomic $(2x + 1)$ and monosomic $(2x - 1)$. Non disjunction could occur in autosomes or in sex chromosomes. Unbalanced gametes can also arise due to failure of pairing between meiotic chromosomes so that they remain as univalents. At metaphase univalents move at random to either poles leading to formation of unbalanced gametes. Trisomers can also originate from translocation heterozygotes, and from ionising radiation when it causes spindle disturbances. In plants trisomy has been extensively studied in Datura by Blakeslee and later in maize, barley, tomato and rye. The Klinefelter's syndrome is known in male XXY sheep which showed testicular hypoplasia but no mental retardation.

Chromosomal evolution of wheat and cotton: Genus Triticale demonstrates the human effort to create a new cereal by crossing wheat and rye. Hexaploid Triticum $(2x + 44)$ is crossed to the diploid secale $(2x = 14)$. The tetraploid hybrid undergoes chromosome duplication to produce the octoploid triticale which combines the characters of wheat and rye. It is resistant to diseases affecting both wheat and rye, and the flour made from its grains has very high protein content. Therefore efforts were made to develop it for commercial use as a crop plant.

A number of cultivated plants are allopolyploids. One of the most important cereals, wheat, represents an allopolyploid series of diploid, tetraploid and hexaploid species. The series is represented by three groups designated *Einkorn* (single seeded), *Emmer* and *Vulgare*.

The *einkorn* group consists of two primitive diploid $(2n = 14)$ species, namely *Triticum monococcum* and the wild *T. boeoticum*. Although not of much use for human consumption because the grain is tightly enclosed in the glumes, the einkorn species are useful as fodder. In some parts of Europe and the Middle East they are used for making dark breads.

The *Emmer* group consists of seven species of tetraploid wheats of which the most important are *Triticum dicoccum* (Persian emmer wheat), and *T. durum*. The origin of emmer wheats took place through hybridisation between an einkorn wheat and a wild species *Aegilops* (goat grass) as explain below.

Triticum boeoticum × *Aegilops speltoides*

(wild einkorn) (goat grass)

$2n = 14$ $2n = 14$

AA BB

↓

Sterile Diploid Hybrid

AB ↓ Allopolyploidy

Triticum dicoccoides

$(2n = 28, \text{AABB})$

fertile

Evolved into

→ *T. dicoccum*

$(2n = 28\ \text{A'A'B'B'})$

Most of the emmer wheats are grown for animal feed, but one *T. durum* has a high gluten content and is particularly useful for making chapatis in India and noodles in western countries.

The *vulgare* group consists of five species of hexaploid wheats $(2n = 42)$ including the economically important bread wheat *T. aestivum*. It is said to have originated through hybridisation between

T. dicoccum (A′A′B′B′) and a different species of goat grass *Aegilops squarrosa* (DD) followed by chromosome doubling. The true bread wheat of today therefore, contains three genomes from three different wheats (A′A′B′B′DD ; $2n = 42$).

Bread wheat has 21 pairs of chromosomes which show an interesting behaviour during meiosis. Normally, chromosomes of wheat species coming from different origins do not pair at meiosis. But the chromosomes belonging to A, B and D genomes that are present in hexaploid bread wheat pair with each other under one condition that chromosome No. 5 of B genome should be absent. Thus chromosome I of A pairs with chromosome I of B; chromosome IA can also pair with ID; and chromosome ID can pair with IB. Similar combinations of pairs exist for other chromosomes of A, B and D genomes. Such chromosomes which belong to different genomes, yet show pairing are called homoeologous.

Apparently, there is a gene on the long arm of chromosome 5B of wheat which suppresses homoeologous pairing. Riley *et al* in 1974 have given the name pairing homoeologous or *ph* to this gene. They have also found that at the beginning of meiosis the positions of the chromosomes on the nuclear membrane are determined by this gene, thereby affecting their pairing behaviour.

In cotton it has been possible to trace the origin of American cottons from hybridisation in the past between New World and Old World cottons. The American cultivated cottons have 52 chromosomes whereas the wild American cottons have only 26 chromosomes. The Indian cultivated cottons also have 26 chromosomes but, these are morphologically different from the 26 chromosomes of the wild New World varieties. It appears that some time in past the American wild cotton must have crossed with the Old World cultivated cotton to produce a hybrid with 13 New World and 13 Old World chromosomes. Chromosome duplication in this hybrid gave rise to the present day tetraploid ($2n = 52$) cultivated cottons in America.

EXERCISE:

1. Explain in detail about deletions and duplications.
3. What is structural change in chromosomes and explain translocations and inversions.
3. What is polyploidy? Explain in detail.
4. Write in detail chromosomal evolution of wheat.
5. Write short notes on
   (i) Deletion
   (ii) Duplication
   (ii) Translocations
   (iv) Reciprocal translocations
   (vi) Inversions (Paracentric)
   (vii) Pericentric inversion
   (viii) Euploidy
   (ix) Allopolyploidy
   (x) Aneuploidy
   (xi) Chromosomal evolution in cotton.

14

Cytoplasmic Inheritance

One of the exceptions to the Mendelian laws of inheritance is the transmission of genetic information from parent to offspring through the cytoplasm and is called extranuclear or cytoplasmic inheritance. The role of cytoplasm in heredity is determined from results of reciprocal crosses in which the sources of male and female gametes are reversed.

In chromosomal inheritance it makes no difference in the transmission of a gene whether it comes from male parent, and identical phenotypes are obtained in reciprocal crosses. However, if there is extranuclear inheritance, the resulting phenotypes in a reciprocal cross are non identical and indicate uniparental transmission, in most cases from the maternal parent. Therefore role of the maternal parent results from the unequal cytoplasm contributions of the male and female parents as most of cytoplasm of the zygote is contributed only from the egg. Very few cases of paternal transmissions are also observed.

Extrachromosomal inheritance is also known as extranuclear inheritance, uniparental inheritance, maternal inheritance, non-Mendelian inheritance or cytoplasmic inheritance through several evidences from a number of criteria as follows, the traits with extranuclear inheritance patterns are identified.

(i) Through reciprocal crosses identical phenotypes are produced.

(ii) Sex-linked inheritance follows a definite pattern which can be easily recognized.

(iii) Plasma genes generally show somatic segregation during mitosis, which is a rare occurrence in case of nuclear genes.

(iv) In certain cases maternal influence is exerted on the offspring due to nutrition, environment or special treatment given to the female parent which affect the cytoplasm of the egg.

(v) Inheritance of some traits are influenced by cytoplasm in the egg, but the expression of the trait is modified by chromosomal genes. Coiling in snails (dextra land sinistral) is good example.

(vi) mRNA which was included in the egg cytoplasm prior to fertilisation can also transmit maternal effects in a developing organism.

(vii) Because of their location outside the nucleus, these genetic factors are referred to as plasmagenes or cytogenes which are associated with chloroplast DNA or mitochondrial DNA.

PLASTID INHERITANCE IN MIRABILIS:

One of the first convincing examples of extranuclear inheritance is the inheritance of leaf variegation in four O'clock (Mirabilis jalapa) plant, which was reported soon after rediscovery of Mendel's laws by Carl Correns in 1909. He found for the first time that plastids could be transmitted to the offspring through the egg cytoplasm. Some of these plants had variegated leaves (leaves with patches of both green and white tissues) in some branches, but it also contained green leaves in some branches and other branches only with white leaves.

In such variegated strain of Mirabilis the cytoplasm of the zygopte contains both green and colourless plastids. During cell division in the zygote while forming embryo, these plastids are unequally distributed. Such an embryo on germination produced a variegated plant with three types of branches – those bearing green foliage, colourless foliage and variegated foliage.

Irrespective of the colour of the branch from which the pollen is used for fertilisation, it was found that seeds produced by green branches gave rise only to green plants; and seeds from colourless branches produced colourless seedlings which did not survive due to lack of chlorophyll. But seeds from the variegated branch could produce three types of progeny; plants with green, colourless or variegated branches. This is because the egg cell of a variegated plant will have both kinds of plastids – green as well as white. During cell division, some cells will receive only green plastids, some only white, and others will receive both green and white plastids. Likewise the progeny of such a branch could be green, white or variegated (Fig. 14.1).

Thus it can be concluded that plastids are cytoplasmic organelles which contain DNA and duplicate themselves independently of nuclear genes, and are distributed more or less equally to daughter cells during mitosis. All the progeny inherit maternal characteristics of foliage through egg cytoplasm.

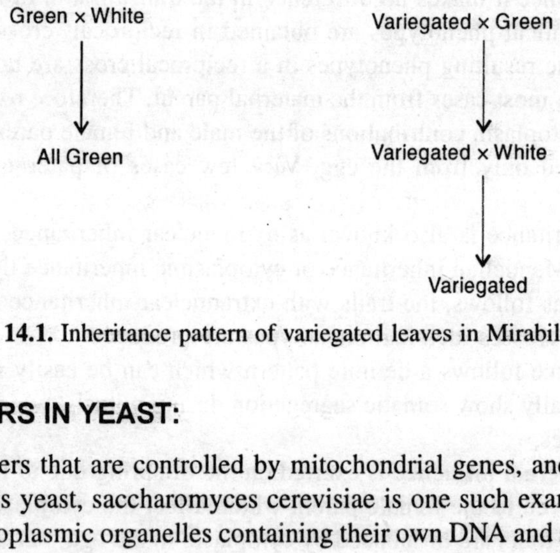

Fig. 14.1. Inheritance pattern of variegated leaves in Mirabilis jalapa

PETITE CHARACTERS IN YEAST:

There are some characters that are controlled by mitochondrial genes, and the inheritance pattern of petite mutants of baker's yeast, saccharomyces cerevisiae is one such example.

Mitochondria are cytoplasmic organelles containing their own DNA and some enzymes that catalyze the process of oxidative phosphorylation. Mitochondria require their own genes and nuclear genes to function normally. Its role in cytoplasmic inheritance was first observed by Ephrussi and his colleagues in 1955. They observed petite mutants (petite = small) in yeast. When a cell divides, approximately equal numbers of mitochondria pass into daughter cells. They can originate from existing mitochondria and can also divide transversely.

As they form smaller colonies on agar compared to normal yeasts, they are referred to as petite mutants. Growth rate of petites is slow due to absence of respiratory enzymes, cytochromes a, b and c, and due to deficiency in some dehydrogenases present in normal mitochondria.

There are three petite strains; the first type called segregational petites when crossed with a normal strain, a 1:1 ratio of normal:petite results after segregation. This suggests Mendelian inheritance and the petite strain has originated due to a mutation in nuclear genes.

The second and third types of petite strains develop when normal yeast cells are treated with the euflavin, ethidium bromide dyes or acidin dyes which are known to intercalate in double stranded DNA. The petites so formed do not segregate regularly when crossed with normal yeast, and are categorised as neutral and suppressive types. When a neutral petite is crossed with a normal type, all the progeny in the next and successive generations are normal and the petite trait never reappears. If a haploid cell of neutral petite strain is mated to a haploid vegetative cell of a normal strain, a diploid zygote is formed. The diploid cells produced by this zygote reproduce asexually, sometimes may even divide meiotically to form spores, but the petite trait is never visible in the progeny. Genetic explanation for this phenomenon is that there is a cytoplasmic factor (p^+) which is present in the normal strains of yeast but absent in petites (p^-).

The suppressive petites when crossed to normal cells of yeast show the petite trait in the progeny but in non-Mendelian ratios. Suppressive petite mutants are found to have mutant DNA in their mitochondria. The mutant mitochondria replicates and transmit the mutant phenotype to the progeny cells.

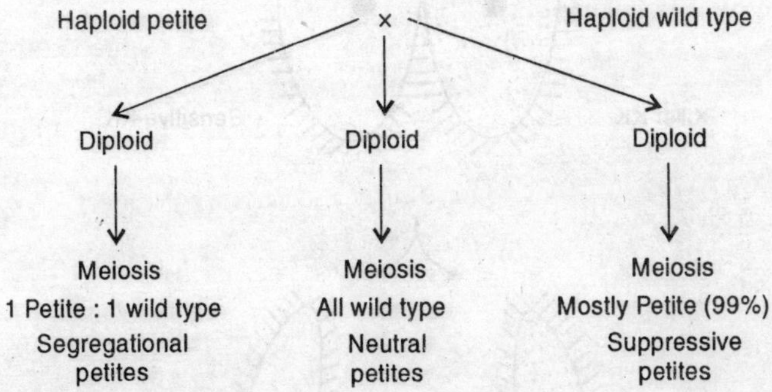

Fig. 14.2. The categories of petite yeasts based on segregation patterns.

KAPPA PARTICLES IN PARAMECIUM

Due to the presence of a cytoplasmic factor called Kappa, in certain strains of Paramecium, showed a killer trait (Someborn, 1938). The killer strain can destroy the sensitive strains growing in culture which do not have kappa by liberating a toxic substance paramecin. However, killer strains are not killed by their own paramecin.

This protozoan has two kinds of nuclei, a small micronucleus and a very large macronucleus which is highly polyploid. The micronucleus behaves according to Mendelian principles while latter behaves irregularly. Paramecium has three modes of reproduction. The first is a simple mitotic division called binary fission. The second method is conjugation. Here two protozoans divide meiotically to form four micronuclei in each cell; out of these, three nuclei degenerate by mitosis to produce two genetically identical haploid nuclei in each paramecium. During conjugation only one of the two haploid nuclei is exchanged through a cytoplasmic bridge formed between the two paramecia. These protozoans then separates as two ex-conjugants.

The third method of reproduction is called autogamy. Here a single paramecium divides meiotically and by the same process that occurs in conjugation, two identical haploid nuclei are formed which fuse to form a diploid organism. As there was no genetic exchange, the diploid paramecium is homozygous.

An important feature of the sensitive strains is that they are not killed by paramecin while they are in the process of conjugation. The two strains can be distinguished morphologically as killers have granular cytoplasm and sensitives are clear.

When a cross is performed between a killer and a sensitive paramecium, there is exchange of genetic material through conjugation. This is followed by separation of the two genetically identical conjugants.

It is found that killer conjugants produce only killer paramecium and the sensitive exconjugants only the sensitive Paramecia. The killer and sensitive traits are obviously not controlled by Mendelian genes (Fig. 14.3).

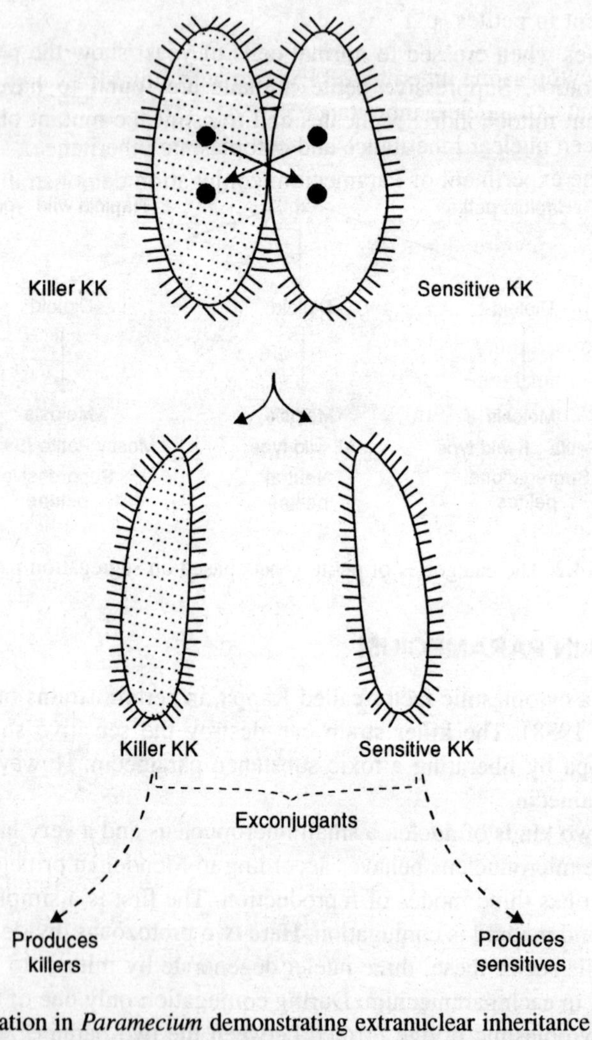

Fig. 14.3. Conjugation in *Paramecium* demonstrating extranuclear inheritance of the killer trait.

By inbreeding the heterozygous (KK) killer exconjugant to another heterozygous killer, it produces three-quarter killer (1KK and 2kk) and one quarter sensitives (kk). But if the sensitive exconjugant (kk) is crossed to another heterozygous sensitive, it results in all sensitive progeny even though their genotypes are in the ratio of 1KK:2Kk:1kk. The results suggest nonchromosomal inheritance of killer trait.

To prove inheritance of killer trait a modified experiment can be conducted. The cross between killer and sensitive has to be prolonged, allowing sufficient time for exchange of cytoplasm. This way, some of the kappa particles could move from killer into the sensitive strain. Entire resulting progeny of such a cross consisted of killers thus confirming the cytoplasmic inheritance of kappa particles.

It was also observed that although kappa particles are transmitted cytoplasmically, yet they need a dominant K gene for maintenance. Kappa are virus like particles about 0.2 microns in diameter and have ability to reproduce independently of the nucleus. As they have their own DNA, can multiply and produce the substance paramecin.

EXERCISE:

1. What is meant by cytoplasmic inheritance? Explain with suitable examples.
2. Write in detail about extranuclear inheritance.
3. Differentiate between nuclear inheritance and cytoplasmic inheritance.
4. Explain in detail the experiment of Paramecium conjugation demonstrating extranuclear inheritance.
5. Write detailed notes on petite characters of yeast, based on segregation patterns.
6. Write short notes on:
   (i) Plastic inheritance
   (ii) Cytoplasmic inheritance
   (iii) Mitochondrial inheritance
   (iv) Kappa particles of Paramecium.

15

Mutations

Mutation is "a sudden, heritable change in phenotype of an individual caused by the simple recombination of genes". The changes include changes in the chromosomes either in structure or in number (chromosomal mutations) and the changes in gene structure of composition (Gene mutations or point mutations).

HISTORICAL BACKGROUND

An English Farmer, 'Seth Wright' discovered first record of a sudden, Heritable change in 1791. He observed a male lamb with unusually short legs, this lead to the source of dominant short-legged trait for the development of 'Ancon' breed of sheep.

Later sin 1900 Hugo de Vries, a Neo-Darwinian observed sudden heritable changes in plant 'Oenothera lamarckiana'. He introduced the term 'MUTATION'.

In 1910 discovery and genetic analysis of WHITE eye mutants of Drosophila by Morgan started the systematic studies of mutations.

Genetic studies relating with spontaneous mutations prompted Geneticists to search for ways of inducing mutations. In this race H.J. Muller (1927) discovered mutagenic action of X-rays in Drosophila, he was awarded the Nobel Prize for physiology and Medicine in 1946. In 1929 Stadler described the mutagenic effects of X-rays in Barley. Later in 1946 Auerbach and Robson discovered mutagenic effects of mustard gas and some other chemical compounds with the induced mutations, first plant breeding programme was initiated in 1929 in Sweden, Germany, USSR and in India.

CLASSIFICATION OF MUTATIONS

There are many criterion to classify the mutations, some of them are given below

1. The Direction of Mutation

Here mutations are divided on the basis of direction of mutational event.

a. Forward Mutation

If mutation occurs from wild type allele to a mutant allele i.e., Forward Mutation.

b. Reverse Mutations

If mutation occurs from a mutant allele to a normal allele i.e., Reverse Mutation.

2. The cause of Mutation

Here mutations are divided on basis of their origin.

a. Spontaneous Mutation

If mutation occurs naturally without any apparent cause i.e., Spontaneous Mutation.

b. Induced Mutation

If mutation occurs in response to a treatment with some mutagens i.e., Induced Mutation.

3. Dominance Relationship

Here mutations are divided on the basis of the role played by the mutant allele.

a. Dominant Mutation

Mutant allele is dominant.

b. Recessive Mutation

Mutant allele is recessive.

c. Co-dominant Mutation:

Mutant allele is Co-dominant.

d. In completely dominant Mutation

Mutant allele is partially dominant.

4. Tissue of Origin

Here mutations are classified on basis of where the mutations originated.

a. Somatic Mutation

Mutation occurs in somatic cell, where the cell is not going to give rise to gametes. If somatic mutation occurs in auxillary bud it is called Bud Mutation.

b. Germinal Mutation

Mutations occur in reproductive tissue, i.e., cells that produce gametes. In sexually reproducing species, only germinal mutations are ordinarily transmitted to next generation. In naturally asexually reproducing species only somatic mutations are transmitted to the progeny.

5. Based on Survival rate

Here mutations are classified on the basis of their effect on survival of the organisms.

a. Lethal Mutations

Mutations which cause death of all individuals carrying these mutations are Lethal Mutations.

b. Sub-Lethal Mutations

Mutations which cause death of the most, but not all individuals that carry in appropriate nature are sub-lethal mutations.

6. Environment influence on Lethal Action

Here Lethal Mutations are further classified.

Some genes are lethal only under certain specific conditions (called Restrictive conditions) while under other conditions (permissive conditions) they do not effect Survival, such genes are known as "conditional Lethals". These are further divided into following categories.

a. Auxotrophic Mutations

These are unable to synthesize a bio-chemical e.g., Vitamins, aminoacids, purines, pyrimidines etc. which the wild type (prototrophic) individuals are capable of synthesizing from relatively simple compounds.

b. Temperature-sensitive Mutations

Some mutants do not survive below (cold-sensitive) or above (Heat-sensitive) certain temperatures, such mutants are called 'Temperature-sensitive mutants (TS mutants).

c. Suppressor-sensitive Mutations

The lethal affect of these mutants are counteracted by another gene, generally a mutant allele called 'Suppressor'.

7. Type of Trait Affected

Here the mutations classified on basis of type of character affected by mutations.

a. Visible or Morphological Mutations

Mutations which change morphological feature of organism, this indicates the presence of mutant allele visually called visible mutations.

b. Biochemical Mutations

Mutations which do not alter the morphological traits, but prevents the production of a biochemical by the organism are Biochemical mutations. Most of these Biochemical mutations are conditional lethal.

8. Intensity of Character Expression

Here visible mutations are further divided on the basis of degree of reduction in expression of characters.

a. Amorphic Mutations

Mutations leads to total loss of expression of traits as they produce totally nonfunctional gene products are 'Amorphic Mutations'.

Eg: White eye in Drosophila.

b. *Hypomorphic Mutations*

Mutations causes partial loss in expression of trait is called 'Hypomorphic Mutations'. Here mutant allele is partly functional.

c. *Isoallelic Mutations*

Mutations which do not effect the intensity of expression of the concerned trait are 'Iso allelic Mutations'. Here mutant allele has same intensity when compared to the wild type allele but mobility is different when subjected to Electrophoresis.

d. *Hypermorphic Mutations*

Mutations that lead to the increased expression than that of wild type are Hypermorphic Mutations.

9. Quantity of Morphological effect produced by Mutation

Here mutations are divided on the basis of Quantum of phenotypic effect changed by mutations.

a. *Macro mutations*

Mutations produce large enough changes in morphology of individuals without any confusion due to environmental effects.

Eg: Short legged sheep.

Macro mutations are observed in qualitative traits governed by polygenes.

10. Effect on the Expression of Neighbouring genes

Here mutations are classified on basis of the effect of mutant alleles on Neighbouring genes.

a. *Polar Mutations*

Mutations which suppress the expression of genes which are located next to them or on one side and also the expression of genes in which they are located.

Usually these are found in prokaryotes.

Eg: In prokaryotes nonsense mutations in a gene interfere with the expression (translation) of not only that gene but also of its immediate neighbouring gene within the same operon.

b. *Non-Polar Mutations*

Mutations which effect the expression of only those genes within which they are located.

11. Mutations on basis of Cytology

Here mutations are classified on their cytological basis.

a. *Chromosomal Mutations*

Mutations which are associated with detectable changes in either chromosome number or structure.

Eg: Bar eye (B) mutation in Drosophila.

b. *Gene Mutations*

Mutations that causes alterations in base sequences are Gene Mutation or Point Mutation.

Both types of mutations involve chromosomes and Nuclear genes which exhibit Mendelian inheritance.

c. Cytoplasmic Mutations

Mutations associated with the changes in chloroplast DNA or Mitochondrial DNA.

Eg: Male sterility due to cytoplasmic mutations is observed in Maize, Jowar etc.

12. Molecular basis

Here Gene mutations are further divided on basis of molecular changes associated with them.

a. Base substitution

If a mutation causes a base to be replaced by another base in the gene it is called 'Base substitution'.

If replacement of purine of purine by pyrimidine by pyrimidine occur it is known as transition.

If replacement of purine by pyrimidine and pyrimidine by purine occur it is known as Transversion.

Eg: For Transversion

GAA   Glustamic acid

↓

GUA   Valine

In the above example II base Adenine replaced by uracil as it is a pyrimidine. This is Transversion.

Eg: For Transition

GAA

↓

GGA

In the above example II base Adenine replaced by Guanine as it is a purine. This is Transition.

b. Deletion Mutation

If mutations cause one or more bases to be deleted from the DNA molecule they are called Deletion Mutations.

c. Addition Mutation

If mutations cause one or more bases to be inserted in the DNA molecule they are called Addition Mutations.

13. Mutations on basis of Replacement of Aminoacid in Polypeptide

Here mutations are classified on the basis of the type of aminoacid replacement they produce in polypeptide chain.

a. Missense Mutation

If a mutation causes replacement of an aminoacid in the polypeptide chain by another aminoacid it is called Missense Mutation. These are produced by base substitutions.

b. Nonsense Mutation

Mutations that generate a codon which does not code for any aminoacid, and subsequently no aminoacid is incorporated in the polypeptide chain and at the nonsense codon translation stops. This results in chain termination. These mutations are called nonsense mutations or chain terminating mutations.

c. Frame-shift Mutations

Mutations that cause shifting of the bases to next position due to either deletion or addition of a base are called Frame-shift Mutations.

MUTAGENESIS

The process of production of mutation is called mutagenesis. This is of two types.

1. Spontaneous mutagenesis.
2. Induced mutagenesis.

1. Spontaneous Mutagenesis

Mutations which are resulted from the either (a) inherent low level of metabolic errors, i.e., mistakes in DNA replication, or (b) they may be actually caused by the mutagenic agents present in the environment of the organisms.

On the basis of chemical considerations it was estimated that DNA polymerase places wrong bases with a frequency of 10^{-15} during DNA replication. But one cycle of proofreading by the $3' \longrightarrow 5'$ exonuclease activity of DNA polymerase, corrects the errors made during replication.

Some of the solar radiations are mutagenic, these include U.V. rays. The mutagenicity of the U.V. rays becomes clear in persons suffering from inherited disease like xeroderma pigmentosum. These patients lack particular enzyme DNA ligase of the DNA repair system, hence they develop mild to severe skin cancer in the areas of body exposed to sunlight.

Rate of Spontaneous Mutations

Usually the rate of Spontaneous Mutations is very low.

In prokaryotes the spontaneous mutational rate is 10^{-8} to 10^{-10}/nucleotide/generation forward mutations.

In Eukaryotes the spontaneous mutation rate is 10^{-7} to 10^{-9}/nucleotide/generation for forward mutations.

The rates of reverse mutations are much lower than the forward mutational rates in prokaryotes.

In Humans

The rate of Spontaneous Mutations are comparatively high in Eukaryotes than prokaryotes. Estimated mutation rates in man appear to the much higher than those for other eukaryotes. Different populations show different mutational rate for some genes.

Different genes of a single species may show markedly different mutation rates.

Eg: The rates of forward mutation of Yellow body colour and to Brown eye colour in Drosophila are estimated as 1.2×10^{-4} and 3×10^{-5} respectively.

INDUCED MUTATIONS

Mutations which are caused by either treating with chemical or physical agent is called Induced Mutations.

Agent which causes mutation is called mutagen.

Eg: α-rays, β-rays, X-rays, γ-rays, Alkylating agents, Base Analogues, Acridine dyes etc.

Treatment with various mutagens increase the mutation rate by many times. Muller observed upto 150 – fold increase in mutational rate following irradiation of Drosophila with X-rays.

Physical Mutagens

Different radiations that have mutagenic effect are known as physical mutagens. These radiations have shorter wave length and higher energy than visible light. These also include the particulate radiations produced by decay of radioisotopes (Eg: ^{14}e, ^{3}e, ^{3}H, ^{35}S etc.)

Radiations are of two types.

1. Ionizing radiation
2. Non-ionizing radiation

Non-ionizing radiation

U.V. rays are the only non-ionizing radiations with Mutagenic properties. The mutagenic effect of U.V. is the consequence of both its direct and indirect effects on DNA.

The direct effect of UV on DNA is of two types.

1. Formation of Pyrimidine dimers and
2. Formation of Pyrimidine hydrates.

Pyrimidines especially Thymine having excited electrons (unpaired electrons) are highly reactive and from pyrimidine dimers with other pyrimidines located next to them in DNA strand.

Pyrimidines especially Thymine having excited electrons (unpaired electrons) are highly reactive and from pyrimidine dimers with other pyrimidines located next to them in DNA strand.

Thymine dimers are most commonly formed, but cytosine dimers and cytosine-thymine dimers are also produced.

DNA polymerase is unable to catalyse the replication of a DNA molecule in region distorted by pyrimidine dimer formation.

Another direct effect of UV on DNA is the Hydration of pyrimidines. Generally cytosine is hydrated, but hydration of Thymine also occurs.

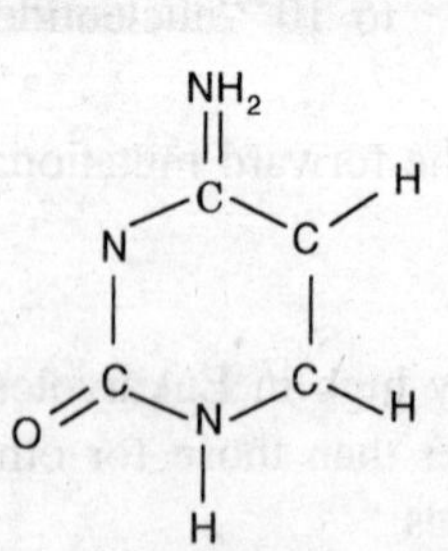

Cytosine

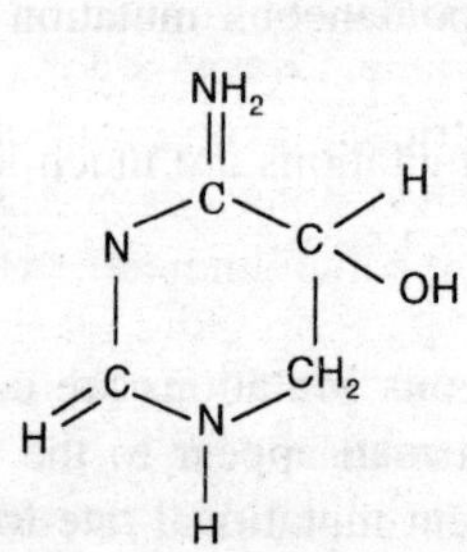

Cytosine Hydrate

Two major effects of Pyrimidine dimers

1. Pyrimidine dimers distort the DNA double helix and interfere with the accurate DNA replication.
2. Errors may occur during excision-repair of pyrimidine dimers.

Indirect Mutagenic effect of UV rays

When normal bacterial cells are placed on to a UV irradiated medium, mutations appear in these cells. This indirect effect is due to organic peroxides and free radicals produced in medium by UV rays. The effect UV on medium is short lived.

Ionizing radiations

Radiation which cause ionization in the atoms present in their path. Two types of ionizing radiations are present.

1. Particulate
2. Non particulate

1. Particulate radiations

They consist of high energy atomic particles generated due to radio active decay of heavier elemcnts.

Slow neutrons are produced when the velocity of fast neutrons is reduced by Graphite or Heavy water.

Both types of Neutrons are highly penetrating in biological tissues.

2. Non-particulate radiations

These are produced by X-rays and γ-rays, which are high energy radiations. The biological effects of X-rays and γ-rays are similar in their physical properties also, but they differ only in their origin.

X-rays produced by X-ray tube. γ-rays produced by the decay of some radioactive isotopes e.g., ^{14}C, ^{60}CO etc.

The genetic effects of radiations may be

1. Direct and
2. Indirect.

The direct effect is produced due to ionization directly in the DNA molecule.

The indirect effect is produced through ionization in molecules other than DNA and is believed to be mediated by the free radical formation.

Free radicals are electrically neutral molecules with unpaired electrons in their outer orbit. They transfer their energy to other molecules making them highly reactive. Free radicals are produced only in those systems that have some amount of water, and almost all biological systems contain water.

$H_2O. \xrightarrow{\text{irradation}} H_2O^+ + e^-$ e^- is ejected from O atom

$H_2O. \xrightarrow{e^-} H_2O.$ e^- is captured by one H atom

In presence of water

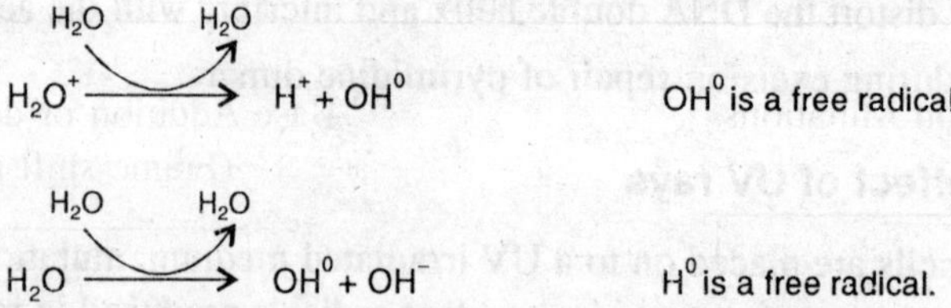

Fig. 15.1. Production of Free radicals.

Detailed discussion of chemical mutagens comes later in this chapter only.

Molecular Basis of Gene Mutation

Mutation in a character would generally be due to a change in aminoacid sequence of protein determining the expression of that trait. The change in aminoacid sequence of this protein would obviously be due to a change in the base sequence of a DNA molecule coding for that protein.

The changes in base sequence is mainly due to either (1) Base substitution or (2) Base addition or deletion.

Base Substitution

When one base is replaced by another base in DNA it is called Base Substitution. It is of 2 types

1. *Transition*

When purine replaces another purine or pyrimidine replaces another pyrimidine it is called transition. This effects only one codon, and so only one aminoacid is altered in the concerned protein.

2. *Transversion*

A pyrimidine replaces a purine and vice-versa. Thus are different types of possibilities of transversions are there.

Base Addition and Deletion

Insertion of one or more bases is Base addition and loss of one or more bases is Base deletion.

If the base which are deleted or removed are the multiple of 3, to many several aminoacids would be either added or deleted from that particular protein such change may/or may not have a striking effect on the activity of the polypeptide.

If the number of bases lost or added are not the multiple of 3, the base sequences of all the codons beyond the point of insertion or deletion are altered. Hence all the codons will code for different aminoacids than before. Thus reading from of these codons are shifted to next place in these mutations. Thus these mutations are called "Frame shift mutations". These mutations change all the aminoacids of the concerned protein which are located beyond point of addition or deletion of bases. Proteins produced from these mutations are non-functional hence such mutations are more deleterious than those produced by base substitutions.

Gene Mutations

- Base Substitution Mutations
  - Transition
    - i. Tautomerization
    - ii. Ionization
    - iii. Base analogues
    - iv. Deamination
  - Transversion
    - i. Alkylating agents
    - ii. Depurination
- Base Addition or deletion mutation (Frame shift mutations)
  - Insertion mutations
  - Deletion mutations

Transition

The transitional substitutions can be created by any of the following ways either during DNA replication or otherwise.

a. Tautomerization

In normal DNA purine, Adenine (A) is linked to the pyrimidine – Thymine (T) by $2H_2$ bonds, while the purine Guanine (G) pairs with pyrimidine, cytosine (C). All these four bases rarely exist in alternate forms called 'Tautomers". These are formed due to "Tautomeric shift", it means the movement of Hydrogen atoms from one position in a purine or pyrimidine to another position. e.g., from a ring nitrogen to a ketone group (C=O) converting it into a Hydroxyl group (–OH) in Tautomerization of Thymine and Guanine or from an Aminogroup ($-NH_2$) to a ring nitrogen imino (–NH) in tanlomerization of cytosine and Adenine. Occurrence of tautomeric shift in a base is called Tautomerization.

In tautomeric state Adenine pairs with the cytosine instead of Thymine, its normal partner and Tautomeric Guanine pairs with Thymine, instead of cytosine. In the same way tautomeric Thymine pairs with normal guanine and cytosine with Adenine. These pairs are known as 'Forbidden base pairs' or "unusual base pairs".

Here we see the example for a simple tautomeric shift. If adenine is in its tautomeric form i.e., iminoform, then the complementary new chain formed from this chain will contain cytosine against adenine (iminoform). In the next step of replication this cytosine pairs with normal Guanine. Thus this entire process produces a substitution of A=T base pair by G≡C pair, similarly a substitution of G≡C by A=T pair can be produced if cytosine appears in its tautomeric form.

Replication of DNA is necessary if the errors produced by tautomeric shifts are to appear in DNA molecules, and therefore classified as COPY ERRORS. This type of tautomeric shifts are not stable because at the next replication the tautomeric base returns to common state and pairs with thymine.

ii. Ionization

Ionizations also causes Transitions. Ionization involves losing of Hydrogen atom from the first Nitrogen of a base.

In ionized state Thymine pairs with Guanine instead of Adenine.
In ionized state Guanine pairs with thymine instead of cytosine.

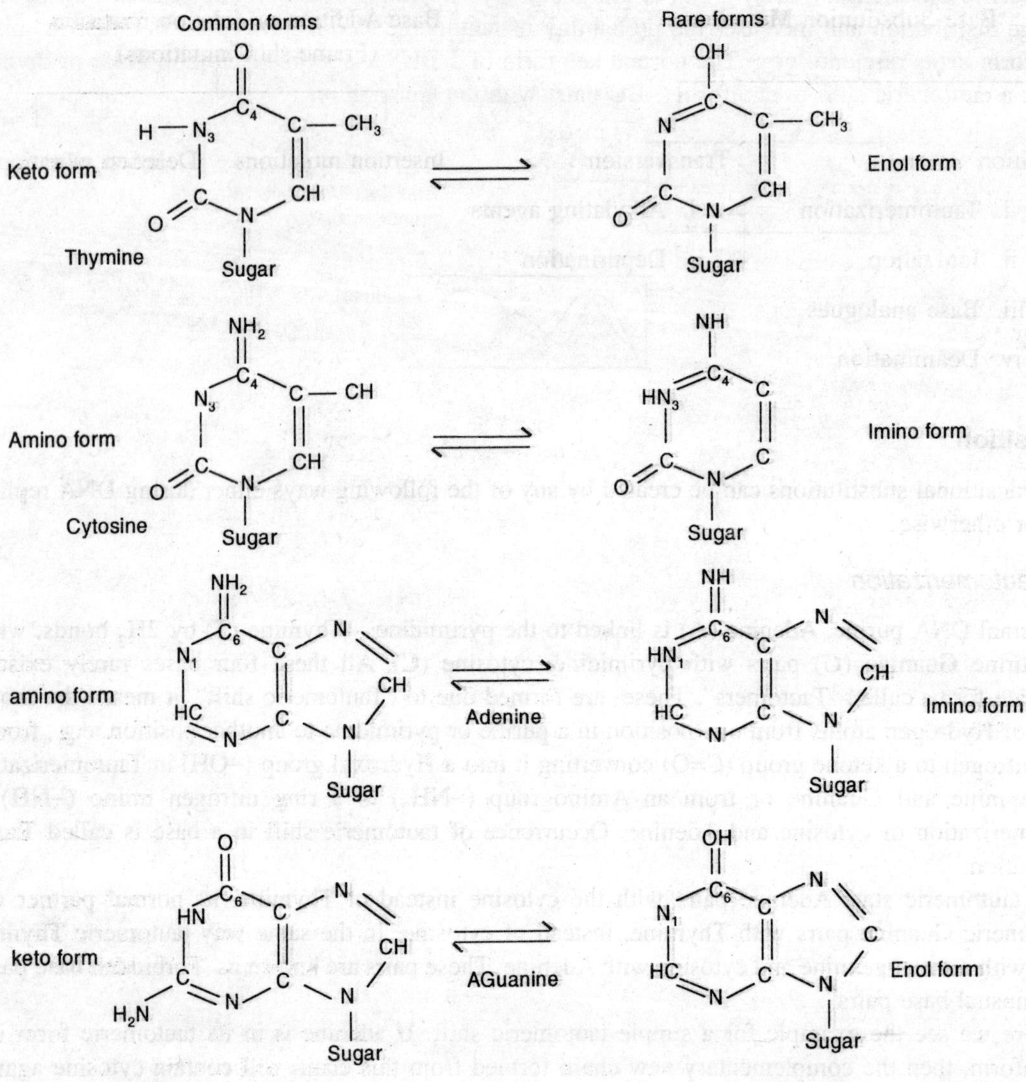

Fig. 15.2

iii. Base Analogues

Chemical compounds which have similar molecular structure as Nitrogen bases of DNA are called 'Base Analogues'. These are usually derivatives of Nitrogenous bases of DNA and occur as natural base analogues as 6-Methyl purine occurs in bacteria, 5-glucosyl hydroxymethyl cytosine, 5-Hydroxymethyl uracil found in some bacteria. The artificial base analogues are 5-Bromouracil (5-Bu) 5-iodouracil (5-IU), 5-Methyl cytosine, 2-Amino purine (2-AP), 2, 6 diaminopurine.

5-IU and 5-BU is base analogue of thymine 5-methyl cytosine is base analogue of cytosine.

a. 5-Bromouracil (5-BU)

The pyrimidine 5-BU is a thymine analogus the Bromine at the 5-position being similar in several respects to the Methyl (-CH_3) group at 5th position in Thymine. The Bromine presence changes the charge distribution and increases the probability of tautomeric shifts. 5-Bu exists in either its normal ketoform or its rare enolform. The normal ketoform of 5-BU pairs with Adenine in place of thymine. After a tautomeric shift to enolform 5-BU pairs with the Guanine.

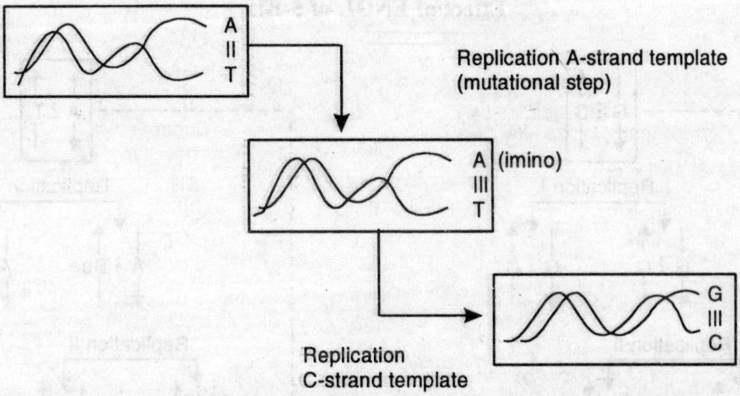

Mechanism of Substitution of A=T pair of G≡C by tautomerization

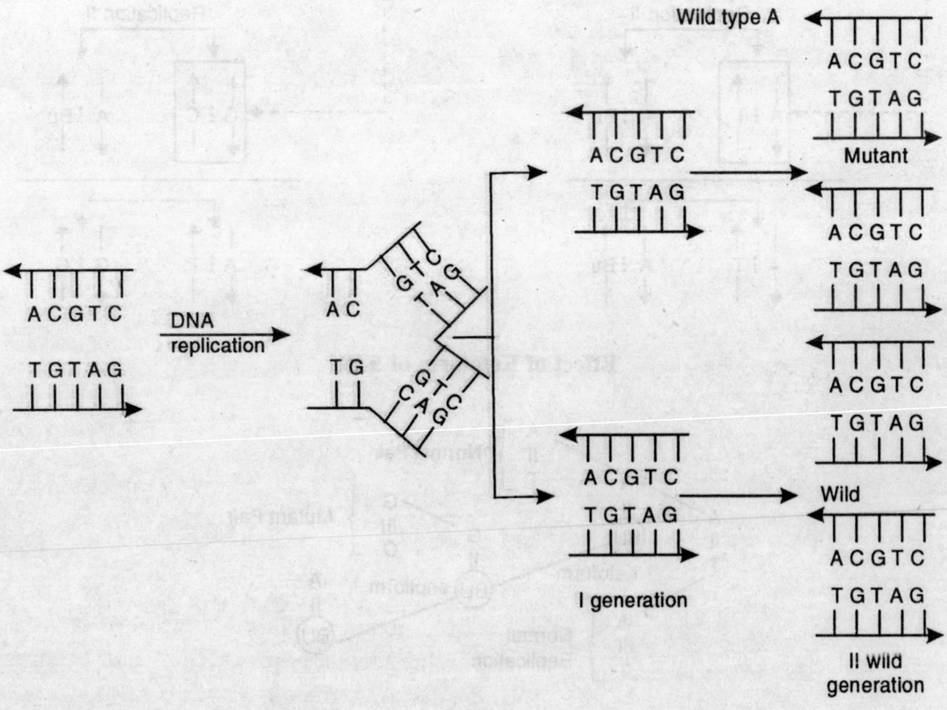

Fig. 15.3

If 5-Bu is in its less frequent enol form as nucleoside triphosphate at the time of incorporation into a nascent strand of DNA, it will be incorporated opposite Guanine in template strand and cause a GC ⟶ AT transition. (Fig. 1) in the same if it is in its frequent ketoform will cause an AT ⟶ GC transition (Fig. 2). Thus 5-BU induces transitions in both the directions AT ⟶ GC. An important consequence of bidirectionality of 5-BU induced transitions is that mutations originally induced with this thymine analog can also be induced to revert with it.

Effect of ENOL of 5-BU

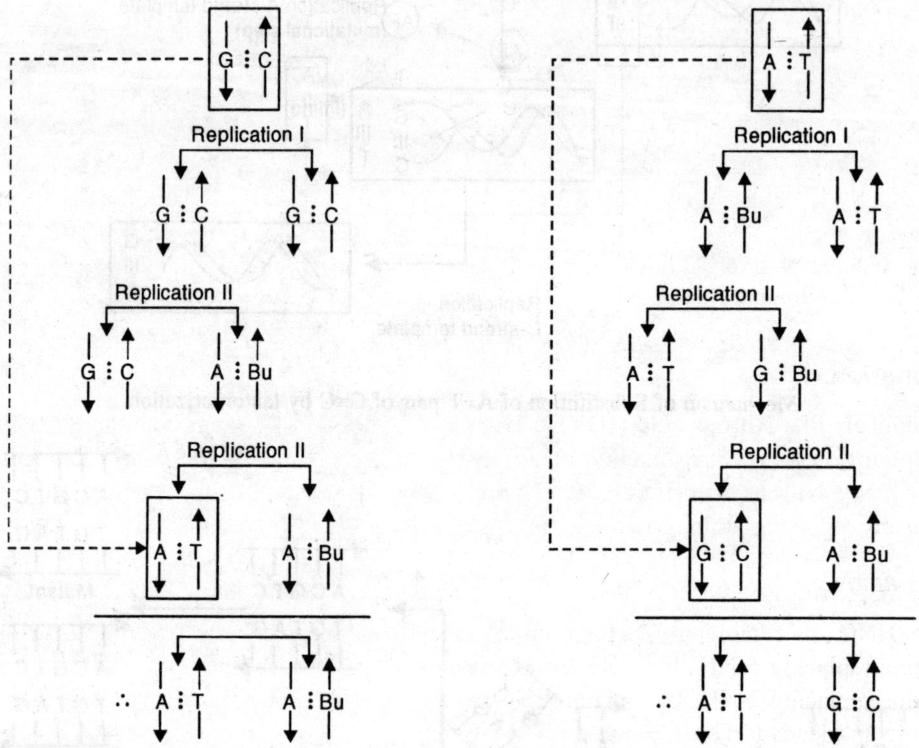

Effect of Ketoform of 5-BU

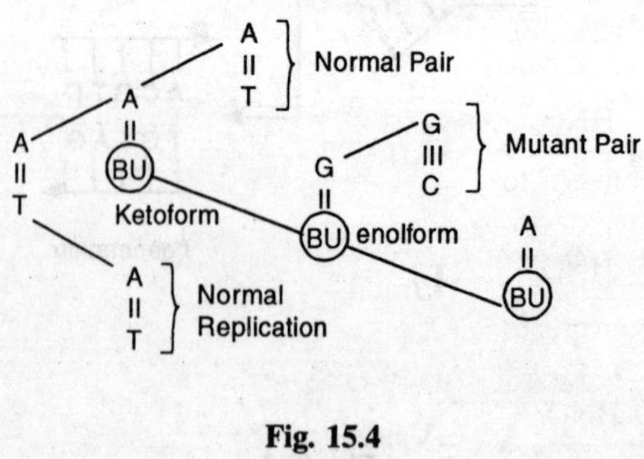

Fig. 15.4

b. Amino purine (2-AP)

2-AP is a artificial base analog for Adenine. It substitutes Adenine and pairs with cytosine. 2-AP specifically induces A : T ⟶ G : C transition.

2-AP can pair with thymine by 2-hydrogen bonds, and with cytosine by a single bond. Pairing between AP and thymine is common because N_2 in cytosine repel each other and cause their separation.

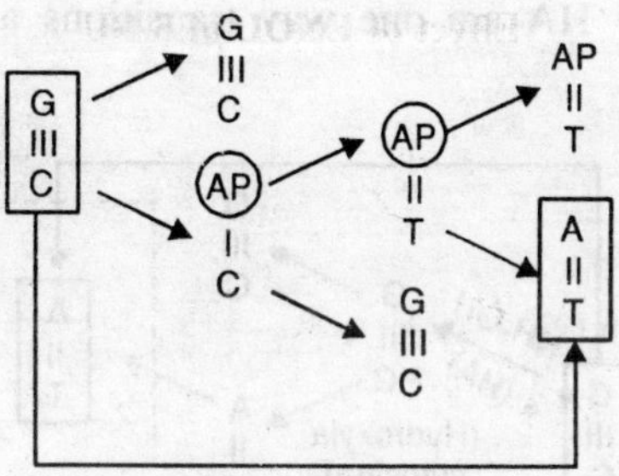

Fig. 15.5. Steps involved in GC ⟶ AT bp transition due to mistake in incorporation of 2-Aminopurine (2-AP).

iv. Deamination

Certain chemicals like Nitrous acid (HNO_2), Hydroxylamine, Ethylmethane Sulfonate (EMS), Ethyl Ethane Sulphonate (EES) Nitrosomethyl urea (NMU), Nitorguanidine (NTG) causes changes in base sequence of DNA. Nitrous acid (HNO_2) and Hydroxylamine causes oxidative deamination of Nitrogen bases. It means replacing aminogroup ($-NH_2$) by Hydroxyl group (–OH).

1. Nitrous acid

In cytosine, HNO_2 by removing 6-NH_2 group and adding –OH group converts it into uracil.

In Adenine replacing 6-NH_2 by –OH group converts it to hypoxanthine.

In Guanine replacing 2-NH_2 by –OH group converts it into xanthine, the newly formed uracil pairs with Adenine and leads to G ≡ C ⟶ A=T transition.

Hypoxanthine pairs with cytosine producing A=T G ≡ C transitions.

Xanthine pairs with cytosine as Guanine does. Hence Deamination of Guamine does not yield mutation.

Thus HNO_2 produces both types of transitions.

Adenine $\xrightarrow[\text{converts to}]{\text{HNO}}$ **H**ypxanthine $\underset{}{\overset{\text{Pairs with}}{\rightleftharpoons}}$ **C**ytosine

Cytosine $\xrightarrow[\text{converts to}]{HNO_2}$ **U**racil $\overset{\text{Pairs with}}{\rightleftharpoons}$ **A**denine

Guanine $\xrightarrow[\text{converts to}]{HNO_2}$ **X**anthine $\overset{\text{Pairs with}}{\rightleftharpoons}$ **C**ytosine

2. Hydroxylamine: (HA = NH_2.OH)

The major mutagenic effect of HA is due to alteration of cytosine. HA causes hydroxylation of cytosine at aminogroup giving rise to Hydroxylcytosine, which then subsequently pairs with Adenine. This is because Hydroxyl – aminogroup is more electronegitive than aminogroup, thus Hydroxylated molecule is more frequently in tautomeric form having H_2 atom in place of N_2 in position-3. This form can't pair with Guanine but with Adenine.

The transitions introduced by HA are one way transitions as these transitions do not undergo reversion.

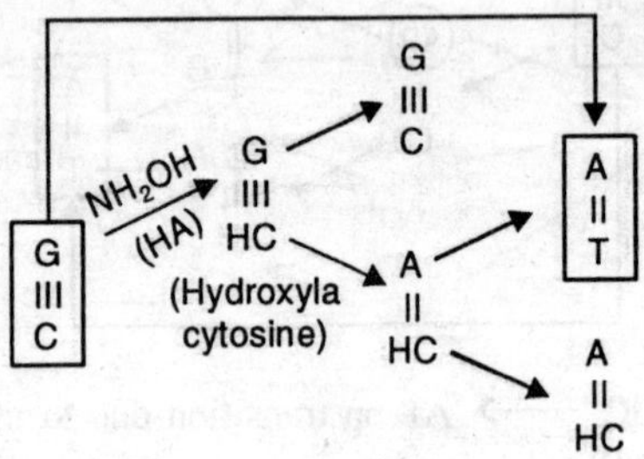

Fig. 15.6

Transversions

Many alkylating agents causes transversions. Many chemicals carry one, two or more alkyl groups in a reactive form. These are called mono, bi or polyfunctional alkylating agents.

Eg: EMS (Ethyl Methane Sulphonate)
EES (Ethyl Ethane Sulphate)
MMS (Methyl Methane Sulphonate)
DES (Diethyl sulphate)
DMS (Dimethyl Sulphate)

Many alkylating agents induce substitutions either by transition or transversion.

One major mechanism of mutagenesis by Alkylating agents involve the transfer of $-CH_3$ or $-C_2H_5$ groups to the bases such that their Base pairing potentials are altered and transitions result.

Eg: In Guanine Ethylation at the 7-N position and at 6-O position are believed to be two effects of EMS. 7-ethylguanine is then believed to base pair with thymine.

The occasional errors occurring in these repair processes may lead to transversions and frame shift mutations in addition to transitions. Some alkylating agents which have two alkyl groups cross link DNA strands and induce chromosomal breaks sand various chromosomal aberrations.

Depurination

Alkylation of Guanine at 7th position gives rise to quaternary N_2 which is unstable. This either hydrolyzes away the alkyl group or else alkylated purine separated from deoxyribose sugar leaving a gap. This removal of a purine from the strand of DNA called Depurination. At the time of replication any of 4 bases can get inserted at this place in the complementary strand.

The depurinated DNA is also more labile and may undergo breakage in backbone soon after this may induce large alternations or may be lethal.

$$\text{Guamine} \xrightarrow{\text{EMS}} \text{M – ethyl Guamine}$$
$$\Vert$$
$$\text{Thymine}$$

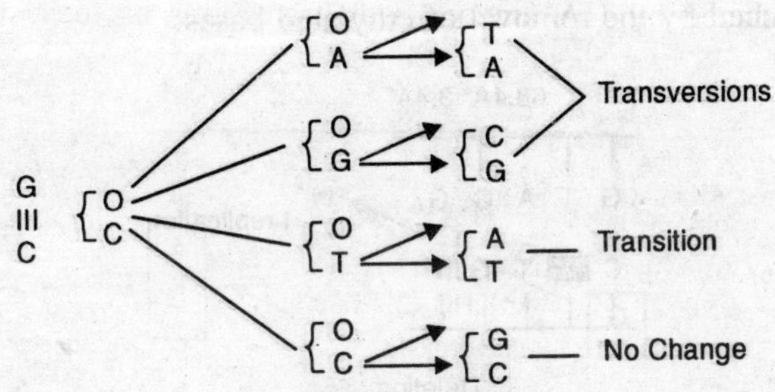

Fig. 15.7. Four different possibilities of Depurination in GC base pair.

Frame Shift Mutations

Addition or deletion of N_2 bases in DNA or mRNA causes Frame shift mutations.

These shift the reading frame of codons from the site of change onward.

Types of frame shift mutations

1. Deletion Mutations

This occurs when one or more bases are deleted from DNA.

2. Insertion Mutations

This occurs when one or more bases are added to DNA molecule at one or more places.

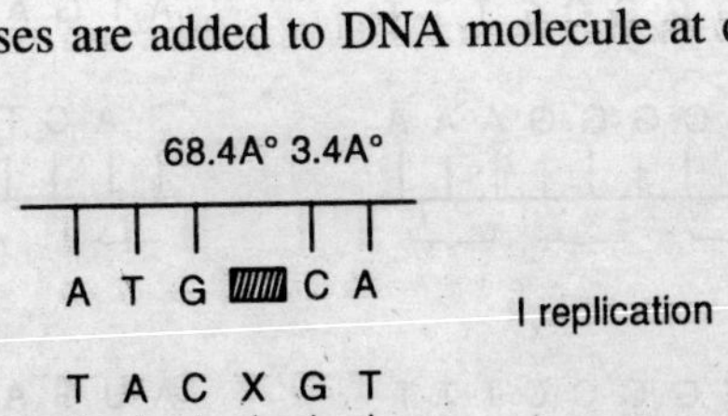

Fig. 15.8. Addition of New bases due to Intercalation

Mechanism

Acridine dyes cause Frame shift mutations.

Eg: Acriflavin, proflavin, series of compounds called ICR-170, ICR-191 and so on.

Acridine dyes like 5-aminoacridine and proflavin intercalate between two stacked base pairs of DNA and increase the distance between them from 3.4 A° to 6.8 A°. At the time of replication, either nitrogenous base pair is inserted or a nitrogenous pair will be lost. Frameshift mutations caused by deletions are also introduced by the removal of ethylated bases.

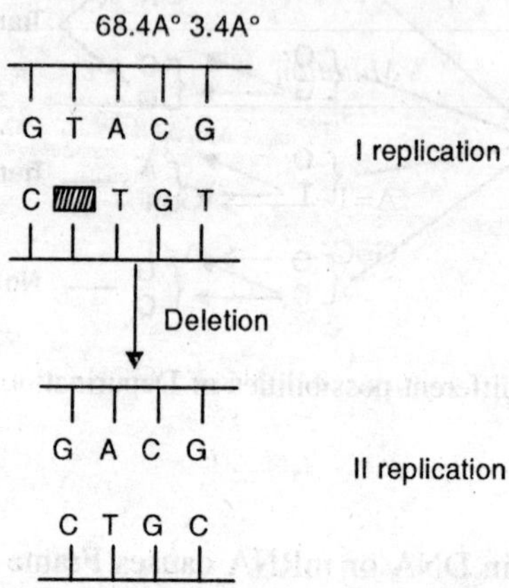

Fig. 15.9. Deletion of bases due to Intercalation

Frameshift mutations cause all the DNA beyond the point of mutations to be misread. Such mutations are lethal (Fig. 15.10). If the deleted segment includes nucleotides in 3 or multiple or 3, there is no frameshift error beyond the site of mutations but protein produced may be defective.

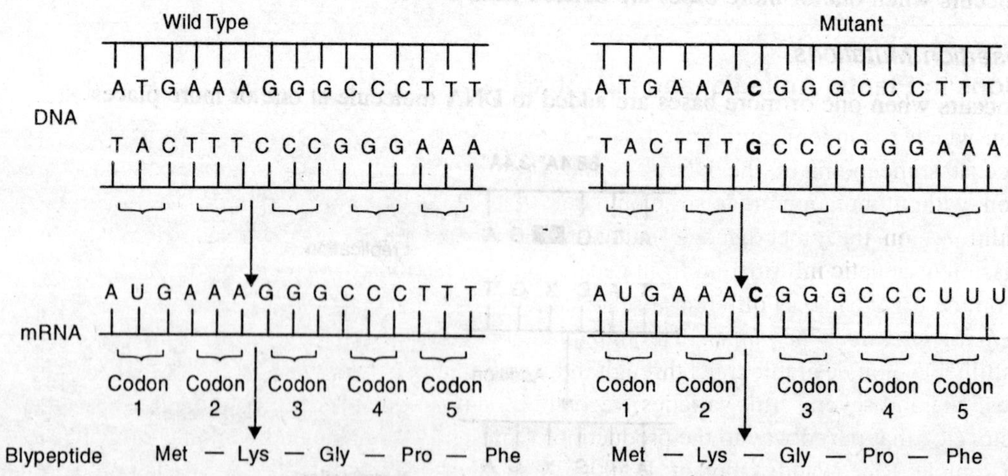

Fig. 15.10. Diagram of a frameshift mutations that results from the addition of a single bp to a structural gene. Mutant gene is produced by the addition of a CG-bp between 6th and 7th position of the wild type gene. This alters the "reading frame" of that portion of gene distal, relative to the direction of transcription and translation to the mutation. As a result, all the codons of mRNA and all the aminoacids of the polypeptide that corresponds to bp-triplets distal to the mutations are altered.

Biological Significance of point Mutations

All point mutations are not lethal. Transition and transversions are relatively begin, because these cause replacement of only one amino acid in the peptide chain coded, which might not produce any significant change. Such mutations are called silent mutations.

Frameshift mutations cause all the DNA beyond the point of mutation to be misread. Such mutations are lethal.

Table: Effect of some chemical mutagens

Mutagen	*Mutation*	*Types of Mutations*
1. Base Analogues		transitions
i. 5-Bromouracil	A=T ⟶ G≡C	transitions
ii. 2-Aminopurine	G≡C ⟶ A=T	transitions
2. Deamination		
iii. Nitrous acid	G≡C ⟶ A=T	transitions
(HNO_2)	G≡C ⟶ G≡C	
iv. Hydroxylamine	G≡C ⟶ A=T	transitions
3. Alkylation	G≡C ⟶ A=T	transitions
(ethylation or methylation)	A=T ⟶ G≡C	transitions
Ethyl Ethane Sulphonate	G≡C ⟶ C-G	transitions
		G≡C ⟶ T=A
		A=T ⟶ C≡G
4. Acridine	Deletion of bases	Deletion Frameshift
	Addition of bases	Addition Frameshift.

Mutations in Plants, Animals, and Microbes for Economic Benefit of Man

Mutation which is random, unidirected, heritable variation caused by an alteration in the nucleotide sequence at some point of the DNA—is the ultimate source of all genetic variation; it leads to evolution, without mutation organisms could not be able to evolve and adapt to respective environmental conditions, on the other hand if mutations occur very frequently, that will totally disrupt the transmission of genetic information from generation to generation. And some mutations cause negative effects on the organisms, as different disease conditions, sometimes turning lethal to that organism.

Beneficial mutations in plants: Through plant breeding and mutagenesis, plant breeders could successfully develop desirable traits through induced mutants in paddy, wheat, barley, soyabeans, oats, tomatoes, gua and several fruit varieties, vegetables, leafy-vegetables and cotton, oil seeds producing plants etc. etc., that paved way to the production of presently available high yielding cultivated strains.

For example: Rice (Paddy) mutant: P-500-23 and other mutants provide high yield. Wheat mutant: Sharbati Sonora, a dwarf variety provide increased yield. Barley mutants have been raised that provide increased yield, with resistance to smut, stiff straw, increased protein content and hullless seed.

Beneficial mutations in animals: Several cultivable fin fishes like Indian major corps—mutants are produced with short-head and large body through mutation breeding.

Short legged sheep which is homozygous for the mutant gene is beneficial as it cannot jump over the fence and run away.

Beneficial mutations in microorganisms: Induced mutations could improve the yield of penicillin by the mould penicillium. Such irradiated and induced over producers of penicillin are proven invaluable in the commercial production of this antibiotic.

Luria and Delbrueck (1946) noticed that when E. coli cells were infected with the bacteriophage T_1, many of the cells lysed releasing progenies of newly synthesized viruses; some bacteria cells, however did not lyse but multiplied and formed colonies—these bacterial cells have developed as phage resistant mutants.

Mutations have been used to understand easily the pathways by which biological processes occur. Metabolism occurs via sequence of enzyme catalysed reactions. Sequence of steps in a pathway can be determined by isolating and studying mutation in the genes coding for that particular enzyme involved. Because an appropriate mutation will be able to eliminate the activity of a single polypeptide; thus mutations provide an extremely powerful method by which dissection of biological processes is possible.

For example:

Precursor (P) ⟶ Intermediate (I) ⟶ Product (M)

enzyme A ↑ gene (A)   enzyme B ↑ gene (B)

Intermediate I is produced from precursor P, by the action of enzyme A which is product of gene A; Intermediate I is rapidly converted to product M by enzyme B which is product of gene B.

Therefore, intermediate I may be present in very low quantities and be very difficult to identify biochemically. Mutant organisms that have mutation in gene B will synthesize either inactive form of enzyme B or no enzyme B at all; so intermediate I will often accumulate to much higher concentration, facilitating its isolation including identification. Similarly, a mutant in gene A could be useful in the identification of precursor P.

Added to this, the functions and mechanisms of action of individual gene products can be often identified by comparative biochemical and biophysical analysis of mutant and wild life.

By involving mutant organisms one can easily separate the pathway of morphogenesis, e.g., T_4 pathway of bacteriophage morphogenesis. Other biological processes which have been successfully dissected by mutational analysis are:

(a) Nitrogen fixation in bacteria
(b) Photosynthetic electron transport chain in chlamydomonas reinbordi and maize.
(c) For understanding differentiation and development in higher plants and animals.
(d) S. Benzer and co-workers used mutant varieties to dissect behavioural learning in Drosophila.

Theoretically under genetic control mutational dissection should be applicable to any process.

EXERCISE:

1. Define mutation and give detailed classification of mutations.
2. In how many ways can a base in DNA be altered to produce a mutation? Narrate with an example of each.

3. In how many different ways can a base in DNA be altered to produce a mutation? Give one example of each.
4. What is induced mutation? Write a note on physical mutagens.
5. Differentiate between
   (a) Spontaneous mutation and induced mutation
   (b) Physical and chemical mutagens
   (c) Transition and transversion.
6. Enumerate significance of mutations. Give some examples of beneficial mutations with reference to human beings.
7. Write short notes on the following:
   (a) Non-ionizing radiation.
   (b) Non-particulate radiation.
   (c) Base addition and deletion.
   (d) Tautomerization.
   (e) Deamination.
   (f) Depurination.
   (g) Point mutations.
   (h) Mutagens.
   (i) Base analogues.